Street-Fighting M

Street-Fighting Mathematics

The Art of Educated Guessing and Opportunistic Problem Solving

Sanjoy Mahajan

Foreword by Carver A. Mead

The MIT Press
Cambridge, Massachusetts
London, England

© 2010 by Sanjoy Mahajan
Foreword © 2010 by Carver A. Mead

Street-Fighting Mathematics: The Art of Educated Guessing and Opportunistic Problem Solving by Sanjoy Mahajan (author), Carver A. Mead (foreword), and MIT Press (publisher) is licensed under the Creative Commons Attribution–Noncommercial–Share Alike 3.0 United States License. A copy of the license is available at http://creativecommons.org/licenses/by-nc-sa/3.0/us/

For information about special quantity discounts, please email special_sales@mitpress.mit.edu

Typeset in Palatino and Euler by the author using ConTEXt and PDFTEX

Library of Congress Cataloging-in-Publication Data

Mahajan, Sanjoy, 1969–
Street-fighting mathematics : the art of educated guessing and opportunistic problem solving / Sanjoy Mahajan ; foreword by Carver A. Mead.
  p. cm.
Includes bibliographical references and index.
ISBN 978-0-262-51429-3 (pbk. : alk. paper) 1. Problem solving. 2. Hypothesis. 3. Estimation theory. I. Title.
QA63.M34 2010
510—dc22

2009028867

Printed and bound in the United States of America
10 9 8 7 6 5 4 3 2 1

For Juliet

Brief contents

	Foreword	xi
	Preface	xiii
1	Dimensions	1
2	Easy cases	13
3	Lumping	31
4	Pictorial proofs	57
5	Taking out the big part	77
6	Analogy	99
	Bibliography	123
	Index	127

Contents

Foreword		xi
Preface		xiii
1	**Dimensions**	**1**
	1.1 Economics: The power of multinational corporations	1
	1.2 Newtonian mechanics: Free fall	3
	1.3 Guessing integrals	7
	1.4 *Summary and further problems*	11
2	**Easy cases**	**13**
	2.1 Gaussian integral revisited	13
	2.2 Plane geometry: The area of an ellipse	16
	2.3 Solid geometry: The volume of a truncated pyramid	17
	2.4 Fluid mechanics: Drag	21
	2.5 *Summary and further problems*	29
3	**Lumping**	**31**
	3.1 Estimating populations: How many babies?	32
	3.2 Estimating integrals	33
	3.3 Estimating derivatives	37
	3.4 Analyzing differential equations: The spring–mass system	42
	3.5 Predicting the period of a pendulum	46
	3.6 *Summary and further problems*	54
4	**Pictorial proofs**	**57**
	4.1 Adding odd numbers	58
	4.2 Arithmetic and geometric means	60
	4.3 Approximating the logarithm	66
	4.4 Bisecting a triangle	70
	4.5 Summing series	73
	4.6 *Summary and further problems*	75

5 Taking out the big part — 77

- 5.1 Multiplication using one and few — 77
- 5.2 Fractional changes and low-entropy expressions — 79
- 5.3 Fractional changes with general exponents — 84
- 5.4 Successive approximation: How deep is the well? — 91
- 5.5 Daunting trigonometric integral — 94
- 5.6 *Summary and further problems* — 97

6 Analogy — 99

- 6.1 Spatial trigonometry: The bond angle in methane — 99
- 6.2 Topology: How many regions? — 103
- 6.3 Operators: Euler–MacLaurin summation — 107
- 6.4 Tangent roots: A daunting transcendental sum — 113
- 6.5 *Bon voyage* — 121

Bibliography — 123

Index — 127

Foreword

Most of us took mathematics courses from mathematicians—Bad Idea!

Mathematicians see mathematics as an area of study in its own right. The rest of us use mathematics as a precise language for expressing relationships among quantities in the real world, and as a tool for deriving quantitative conclusions from these relationships. For that purpose, mathematics courses, as they are taught today, are seldom helpful and are often downright destructive.

As a student, I promised myself that if I ever became a teacher, I would never put a student through that kind of teaching. I have spent my life trying to find direct and transparent ways of seeing reality and trying to express these insights quantitatively, and I have never knowingly broken my promise.

With rare exceptions, the mathematics that I have found most useful was learned in science and engineering classes, on my own, or from this book. *Street-Fighting Mathematics* is a breath of fresh air. Sanjoy Mahajan teaches us, in the most friendly way, tools that work in the real world. Just when we think that a topic is obvious, he brings us up to another level. My personal favorite is the approach to the Navier–Stokes equations: so nasty that I would never even attempt a solution. But he leads us through one, gleaning gems of insight along the way.

In this little book are insights for every one of us. I have personally adopted several of the techniques that you will find here. I recommend it highly to every one of you.

—Carver Mead

Preface

Too much mathematical rigor teaches *rigor mortis:* the fear of making an unjustified leap even when it lands on a correct result. Instead of paralysis, have courage—shoot first and ask questions later. Although unwise as public policy, it is a valuable problem-solving philosophy, and it is the theme of this book: how to guess answers without a proof or an exact calculation.

Educated guessing and opportunistic problem solving require a toolbox. A tool, to paraphrase George Polya, is a trick I use twice. This book builds, sharpens, and demonstrates tools useful across diverse fields of human knowledge. The diverse examples help separate the tool—the general principle—from the particular applications so that you can grasp and transfer the tool to problems of particular interest to you.

The examples used to teach the tools include guessing integrals without integrating, refuting a common argument in the media, extracting physical properties from nonlinear differential equations, estimating drag forces without solving the Navier–Stokes equations, finding the shortest path that bisects a triangle, guessing bond angles, and summing infinite series whose every term is unknown and transcendental.

This book complements works such as *How to Solve It* [37], *Mathematics and Plausible Reasoning* [35, 36], and *The Art and Craft of Problem Solving* [49]. They teach how to solve exactly stated problems exactly, whereas life often hands us partly defined problems needing only moderately accurate solutions. A calculation accurate only to a factor of 2 may show that a proposed bridge would never be built or a circuit could never work. The effort saved by not doing the precise analysis can be spent inventing promising new designs.

This book grew out of a short course of the same name that I taught for several years at MIT. The students varied widely in experience: from first-year undergraduates to graduate students ready for careers in research and teaching. The students also varied widely in specialization:

from physics, mathematics, and management to electrical engineering, computer science, and biology. Despite or because of the diversity, the students seemed to benefit from the set of tools and to enjoy the diversity of illustrations and applications. I wish the same for you.

How to use this book

Aristotle was tutor to the young Alexander of Macedon (later, Alexander the Great). As ancient royalty knew, a skilled and knowledgeable tutor is the most effective teacher [8]. A skilled tutor makes few statements and asks many questions, for she knows that questioning, wondering, and discussing promote long-lasting learning. Therefore, questions of two types are interspersed through the book.

Questions marked with a ▶ *in the margin:* These questions are what a tutor might ask you during a tutorial, and ask you to work out the next steps in an analysis. They are answered in the subsequent text, where you can check your solutions and my analysis.

Numbered problems: These problems, marked with a shaded background, are what a tutor might give you to take home after a tutorial. They ask you to practice the tool, to extend an example, to use several tools together, and even to resolve (apparent) paradoxes.

Try many questions of both types!

Copyright license

This book is licensed under the same license as MIT's OpenCourseWare: a Creative Commons Attribution-Noncommercial-Share Alike license. The publisher and I encourage you to use, improve, and share the work noncommercially, and we will gladly receive any corrections and suggestions.

Acknowledgments

I gratefully thank the following individuals and organizations.

For the title: Carl Moyer.

For editorial guidance: Katherine Almeida and Robert Prior.

For sweeping, thorough reviews of the manuscript: Michael Gottlieb, David Hogg, David MacKay, and Carver Mead.

For being inspiring teachers: John Allman, Arthur Eisenkraft, Peter Goldreich, John Hopfield, Jon Kettenring, Geoffrey Lloyd, Donald Knuth, Carver Mead, David Middlebrook, Sterl Phinney, and Edwin Taylor.

For many valuable suggestions and discussions: Shehu Abdussalam, Daniel Corbett, Dennis Freeman, Michael Godfrey, Hans Hagen, Jozef Hanc, Taco Hoekwater, Stephen Hou, Kayla Jacobs, Aditya Mahajan, Haynes Miller, Elisabeth Moyer, Hubert Pham, Benjamin Rapoport, Rahul Sarpeshkar, Madeleine Sheldon-Dante, Edwin Taylor, Tadashi Tokieda, Mark Warner, and Joshua Zucker.

For advice on the process of writing: Carver Mead and Hillary Rettig.

For advice on the book design: Yasuyo Iguchi.

For advice on free licensing: Daniel Ravicher and Richard Stallman.

For the free software used for calculations: Fredrik Johansson (mpmath), the Maxima project, and the Python community.

For the free software used for typesetting: Hans Hagen and Taco Hoekwater (ConTEXt); Han The Thanh (PDFTEX); Donald Knuth (TEX); John Hobby (MetaPost); John Bowman, Andy Hammerlindl, and Tom Prince (Asymptote); Matt Mackall (Mercurial); Richard Stallman (Emacs); and the Debian GNU/Linux project.

For supporting my work in science and mathematics teaching: The Whitaker Foundation in Biomedical Engineering; the Hertz Foundation; the Master and Fellows of Corpus Christi College, Cambridge; the MIT Teaching and Learning Laboratory and the Office of the Dean for Undergraduate Education; and especially Roger Baker, John Williams, and the Trustees of the Gatsby Charitable Foundation.

Bon voyage

As our first tool, let's welcome a visitor from physics and engineering: the method of dimensional analysis.

1
Dimensions

1.1	Economics: The power of multinational corporations	1
1.2	Newtonian mechanics: Free fall	3
1.3	Guessing integrals	7
1.4	*Summary and further problems*	11

Our first street-fighting tool is dimensional analysis or, when abbreviated, dimensions. To show its diversity of application, the tool is introduced with an economics example and sharpened on examples from Newtonian mechanics and integral calculus.

1.1 Economics: The power of multinational corporations

Critics of globalization often make the following comparison [25] to prove the excessive power of multinational corporations:

> In Nigeria, a relatively economically strong country, the GDP [gross domestic product] is $99 billion. The net worth of Exxon is $119 billion. "When multinationals have a net worth higher than the GDP of the country in which they operate, what kind of power relationship are we talking about?" asks Laura Morosini.

Before continuing, explore the following question:

▶ *What is the most egregious fault in the comparison between Exxon and Nigeria?*

The field is competitive, but one fault stands out. It becomes evident after unpacking the meaning of GDP. A GDP of $99 billion is shorthand for a monetary flow of $99 billion per year. A year, which is the time for the earth to travel around the sun, is an astronomical phenomenon that has been arbitrarily chosen for measuring a social phenomenon—namely, monetary flow.

Suppose instead that economists had chosen the decade as the unit of time for measuring GDP. Then Nigeria's GDP (assuming the flow remains steady from year to year) would be roughly $1 trillion per decade and be reported as $1 trillion. Now Nigeria towers over Exxon, whose puny assets are a mere one-tenth of Nigeria's GDP. To deduce the opposite conclusion, suppose the week were the unit of time for measuring GDP. Nigeria's GDP becomes $2 billion per week, reported as $2 billion. Now puny Nigeria stands helpless before the mighty Exxon, 50-fold larger than Nigeria.

A valid economic argument cannot reach a conclusion that depends on the astronomical phenomenon chosen to measure time. The mistake lies in comparing incomparable quantities. Net worth is an amount: It has dimensions of money and is typically measured in units of dollars. GDP, however, is a flow or rate: It has dimensions of money per time and typical units of dollars per year. (A dimension is general and independent of the system of measurement, whereas the unit is how that dimension is measured in a particular system.) Comparing net worth to GDP compares a monetary amount to a monetary flow. Because their dimensions differ, the comparison is a category mistake [39] and is therefore guaranteed to generate nonsense.

> **Problem 1.1 Units or dimensions?**
> Are meters, kilograms, and seconds units or dimensions? What about energy, charge, power, and force?

A similarly flawed comparison is length per time (speed) versus length: "I walk $1.5\,\mathrm{m\,s^{-1}}$—much smaller than the Empire State building in New York, which is 300 m high." It is nonsense. To produce the opposite but still nonsense conclusion, measure time in hours: "I walk 5400 m/hr—much larger than the Empire State building, which is 300 m high."

I often see comparisons of corporate and national power similar to our Nigeria–Exxon example. I once wrote to one author explaining that I sympathized with his conclusion but that his argument contained a fatal dimensional mistake. He replied that I had made an interesting point but that the numerical comparison showing the country's weakness was stronger as he had written it, so he was leaving it unchanged!

A dimensionally valid comparison would compare like with like: either Nigeria's GDP with Exxon's revenues, or Exxon's net worth with Nigeria's net worth. Because net worths of countries are not often tabulated, whereas corporate revenues are widely available, try comparing Exxon's annual revenues with Nigeria's GDP. By 2006, Exxon had become Exxon Mobil with annual revenues of roughly $350 billion—almost twice Nigeria's 2006 GDP of $200 billion. This valid comparison is stronger than the flawed one, so retaining the flawed comparison was not even expedient!

That compared quantities must have identical dimensions is a necessary condition for making valid comparisons, but it is not sufficient. A costly illustration is the 1999 Mars Climate Orbiter (MCO), which crashed into the surface of Mars rather than slipping into orbit around it. The cause, according to the Mishap Investigation Board (MIB), was a mismatch between English and metric units [26, p. 6]:

> The MCO MIB has determined that the root cause for the loss of the MCO spacecraft was the failure to use metric units in the coding of a ground software file, Small Forces, used in trajectory models. Specifically, thruster performance data in English units instead of metric units was used in the software application code titled SM_FORCES (small forces). A file called Angular Momentum Desaturation (AMD) contained the output data from the SM_FORCES software. The data in the AMD file was required to be in metric units per existing software interface documentation, and the trajectory modelers assumed the data was provided in metric units per the requirements.

Make sure to mind your dimensions and units.

> **Problem 1.2   Finding bad comparisons**
> Look for everyday comparisons—for example, on the news, in the newspaper, or on the Internet—that are dimensionally faulty.

1.2 Newtonian mechanics: Free fall

Dimensions are useful not just to debunk incorrect arguments but also to generate correct ones. To do so, the quantities in a problem need to have dimensions. As a contrary example showing what not to do, here is how many calculus textbooks introduce a classic problem in motion:

> A ball initially at rest falls from a height of h **feet** and hits the ground at a speed of v **feet per second**. Find v assuming a gravitational acceleration of g **feet per second squared** and neglecting air resistance.

The units such as feet or feet per second are highlighted in boldface because their inclusion is so frequent as to otherwise escape notice, and their inclusion creates a significant problem. Because the height is h feet, the variable h does not contain the units of height: h is therefore dimensionless. (For h to have dimensions, the problem would instead state simply that the ball falls from a height h; then the dimension of length would belong to h.) A similar explicit specification of units means that the variables g and v are also dimensionless. Because g, h, and v are dimensionless, any comparison of v with quantities derived from g and h is a comparison between dimensionless quantities. It is therefore always dimensionally valid, so dimensional analysis cannot help us guess the impact speed.

Giving up the valuable tool of dimensions is like fighting with one hand tied behind our back. Thereby constrained, we must instead solve the following differential equation with initial conditions:

$$\frac{d^2y}{dt^2} = -g, \text{ with } y(0) = h \text{ and } dy/dt = 0 \text{ at } t = 0, \tag{1.1}$$

where $y(t)$ is the ball's height, dy/dt is the ball's velocity, and g is the gravitational acceleration.

> **Problem 1.3 Calculus solution**
> Use calculus to show that the free-fall differential equation $d^2y/dt^2 = -g$ with initial conditions $y(0) = h$ and $dy/dt = 0$ at $t = 0$ has the following solution:
>
> $$\frac{dy}{dt} = -gt \quad \text{and} \quad y = -\frac{1}{2}gt^2 + h. \tag{1.2}$$

▶ *Using the solutions for the ball's position and velocity in Problem 1.3, what is the impact speed?*

When $y(t) = 0$, the ball meets the ground. Thus the impact time t_0 is $\sqrt{2h/g}$. The impact velocity is $-gt_0$ or $-\sqrt{2gh}$. Therefore the impact speed (the unsigned velocity) is $\sqrt{2gh}$.

This analysis invites several algebra mistakes: forgetting to take a square root when solving for t_0, or dividing rather than multiplying by g when finding the impact velocity. Practice—in other words, making and correcting many mistakes—reduces their prevalence in simple problems, but complex problems with many steps remain minefields. We would like less error-prone methods.

1.2 Newtonian mechanics: Free fall

One robust alternative is the method of dimensional analysis. But this tool requires that at least one quantity among v, g, and h have dimensions. Otherwise, every candidate impact speed, no matter how absurd, equates dimensionless quantities and therefore has valid dimensions.

Therefore, let's restate the free-fall problem so that the quantities retain their dimensions:

> A ball initially at rest falls from a height h and hits the ground at speed v. Find v assuming a gravitational acceleration g and neglecting air resistance.

The restatement is, first, shorter and crisper than the original phrasing:

> A ball initially at rest falls from a height of h feet and hits the ground at a speed of v feet per second. Find v assuming a gravitational acceleration of g feet per second squared and neglecting air resistance.

Second, the restatement is more general. It makes no assumption about the system of units, so it is useful even if meters, cubits, or furlongs are the unit of length. Most importantly, the restatement gives dimensions to h, g, and v. Their dimensions will almost uniquely determine the impact speed—without our needing to solve a differential equation.

The dimensions of height h are simply length or, for short, L. The dimensions of gravitational acceleration g are length per time squared or LT^{-2}, where T represents the dimension of time. A speed has dimensions of LT^{-1}, so v is a function of g and h with dimensions of LT^{-1}.

> **Problem 1.4 Dimensions of familiar quantities**
> In terms of the basic dimensions length L, mass M, and time T, what are the dimensions of energy, power, and torque?

▶ *What combination of g and h has dimensions of speed?*

The combination $\sqrt{gh}$ has dimensions of speed.

$$\left(\underbrace{LT^{-2}}_{g} \times \underbrace{L}_{h}\right)^{1/2} = \sqrt{L^2 T^{-2}} = \underbrace{LT^{-1}}_{\text{speed}}. \tag{1.3}$$

▶ *Is $\sqrt{gh}$ the only combination of g and h with dimensions of speed?*

In order to decide whether $\sqrt{gh}$ is the only possibility, use constraint propagation [43]. The strongest constraint is that the combination of g and h, being a speed, should have dimensions of inverse time (T^{-1}). Because h contains no dimensions of time, it cannot help construct T^{-1}. Because g contains T^{-2}, the T^{-1} must come from $\sqrt{g}$. The second constraint is that the combination contain L^1. The $\sqrt{g}$ already contributes $L^{1/2}$, so the missing $L^{1/2}$ must come from $\sqrt{h}$. The two constraints thereby determine uniquely how g and h appear in the impact speed v.

The exact expression for v is, however, not unique. It could be $\sqrt{gh}$, $\sqrt{2gh}$, or, in general, $\sqrt{gh} \times$ dimensionless constant. The idiom of multiplication by a dimensionless constant occurs frequently and deserves a compact notation akin to the equals sign:

$$v \sim \sqrt{gh}. \tag{1.4}$$

Including this $\sim$ notation, we have several species of equality:

$$\begin{aligned} &\propto && \text{equality except perhaps for a factor with dimensions,} \\ &\sim && \text{equality except perhaps for a factor without dimensions,} \\ &\approx && \text{equality except perhaps for a factor close to 1.} \end{aligned} \tag{1.5}$$

The exact impact speed is $\sqrt{2gh}$, so the dimensions result $\sqrt{gh}$ contains the entire functional dependence! It lacks only the dimensionless factor $\sqrt{2}$, and these factors are often unimportant. In this example, the height might vary from a few centimeters (a flea hopping) to a few meters (a cat jumping from a ledge). The factor-of-100 variation in height contributes a factor-of-10 variation in impact speed. Similarly, the gravitational acceleration might vary from $0.27\,\mathrm{m\,s^{-2}}$ (on the asteroid Ceres) to $25\,\mathrm{m\,s^{-2}}$ (on Jupiter). The factor-of-100 variation in g contributes another factor-of-10 variation in impact speed. Much variation in the impact speed, therefore, comes not from the dimensionless factor $\sqrt{2}$ but rather from the symbolic factors—which are computed exactly by dimensional analysis.

Furthermore, not calculating the exact answer can be an advantage. Exact answers have all factors and terms, permitting less important information, such as the dimensionless factor $\sqrt{2}$, to obscure important information such as $\sqrt{gh}$. As William James advised, "The art of being wise is the art of knowing what to overlook" [19, Chapter 22].

> **Problem 1.5 Vertical throw**
> You throw a ball directly upward with speed v_0. Use dimensional analysis to estimate how long the ball takes to return to your hand (neglecting air resistance). Then find the exact time by solving the free-fall differential equation. What dimensionless factor was missing from the dimensional-analysis result?

1.3 Guessing integrals

The analysis of free fall (Section 1.2) shows the value of not separating dimensioned quantities from their units. However, what if the quantities are dimensionless, such as the 5 and x in the following Gaussian integral:

$$\int_{-\infty}^{\infty} e^{-5x^2}\, dx\,? \tag{1.6}$$

Alternatively, the dimensions might be unspecified—a common case in mathematics because it is a universal language. For example, probability theory uses the Gaussian integral

$$\int_{x_1}^{x_2} e^{-x^2/2\sigma^2}\, dx, \tag{1.7}$$

where x could be height, detector error, or much else. Thermal physics uses the similar integral

$$\int e^{-\frac{1}{2}mv^2/kT}\, dv, \tag{1.8}$$

where v is a molecular speed. Mathematics, as the common language, studies their common form $\int e^{-\alpha x^2}$ without specifying the dimensions of α and x. The lack of specificity gives mathematics its power of abstraction, but it makes using dimensional analysis difficult.

▶ *How can dimensional analysis be applied without losing the benefits of mathematical abstraction?*

The answer is to find the quantities with unspecified dimensions and then to assign them a consistent set of dimensions. To illustrate the approach, let's apply it to the general definite Gaussian integral

$$\int_{-\infty}^{\infty} e^{-\alpha x^2}\, dx. \tag{1.9}$$

Unlike its specific cousin with $\alpha = 5$, which is the integral $\int_{-\infty}^{\infty} e^{-5x^2}\, dx$, the general form does not specify the dimensions of x or α—and that openness provides the freedom needed to use the method of dimensional analysis.

The method requires that any equation be dimensionally valid. Thus, in the following equation, the left and right sides must have identical dimensions:

$$\int_{-\infty}^{\infty} e^{-\alpha x^2} \, dx = \text{something}. \tag{1.10}$$

▶ *Is the right side a function of x? Is it a function of α? Does it contain a constant of integration?*

The left side contains no symbolic quantities other than x and α. But x is the integration variable and the integral is over a definite range, so x disappears upon integration (and no constant of integration appears). Therefore, the right side—the "something"—is a function only of α. In symbols,

$$\int_{-\infty}^{\infty} e^{-\alpha x^2} \, dx = f(\alpha). \tag{1.11}$$

The function f might include dimensionless numbers such as $2/3$ or $\sqrt{\pi}$, but α is its only input with dimensions.

For the equation to be dimensionally valid, the integral must have the same dimensions as $f(\alpha)$, and the dimensions of $f(\alpha)$ depend on the dimensions of α. Accordingly, the dimensional-analysis procedure has the following three steps:

Step 1. Assign dimensions to α (Section 1.3.1).

Step 2. Find the dimensions of the integral (Section 1.3.2).

Step 3. Make an $f(\alpha)$ with those dimensions (Section 1.3.3).

1.3.1 Assigning dimensions to α

The parameter α appears in an exponent. An exponent specifies how many times to multiply a quantity by itself. For example, here is 2^n:

$$2^n = \underbrace{2 \times 2 \times \cdots \times 2}_{n \text{ terms}}. \tag{1.12}$$

The notion of "how many times" is a pure number, so an exponent is dimensionless.

Hence the exponent $-\alpha x^2$ in the Gaussian integral is dimensionless. For convenience, denote the dimensions of α by $[\alpha]$ and of x by $[x]$. Then

$$[\alpha][x]^2 = 1, \tag{1.13}$$

1.3 Guessing integrals

or
$$[\alpha] = [x]^{-2}. \tag{1.14}$$

This conclusion is useful, but continuing to use unspecified but general dimensions requires lots of notation, and the notation risks burying the reasoning.

The simplest alternative is to make x dimensionless. That choice makes α and $f(\alpha)$ dimensionless, so any candidate for $f(\alpha)$ would be dimensionally valid, making dimensional analysis again useless. The simplest effective alternative is to give x simple dimensions—for example, length. (This choice is natural if you imagine the x axis lying on the floor.) Then $[\alpha] = L^{-2}$.

1.3.2 Dimensions of the integral

The assignments $[x] = L$ and $[\alpha] = L^{-2}$ determine the dimensions of the Gaussian integral. Here is the integral again:

$$\int_{-\infty}^{\infty} e^{-\alpha x^2} \, dx. \tag{1.15}$$

The dimensions of an integral depend on the dimensions of its three pieces: the integral sign $\int$, the integrand $e^{-\alpha x^2}$, and the differential dx.

The integral sign originated as an elongated S for *Summe*, the German word for sum. In a valid sum, all terms have identical dimensions: The fundamental principle of dimensions requires that apples be added only to apples. For the same reason, the entire sum has the same dimensions as any term. Thus, the summation sign—and therefore the integration sign—do not affect dimensions: The integral sign is dimensionless.

> **Problem 1.6 Integrating velocity**
> Position is the integral of velocity. However, position and velocity have different dimensions. How is this difference consistent with the conclusion that the integration sign is dimensionless?

Because the integration sign is dimensionless, the dimensions of the integral are the dimensions of the exponential factor $e^{-\alpha x^2}$ multiplied by the dimensions of dx. The exponential, despite its fierce exponent $-\alpha x^2$, is merely several copies of e multiplied together. Because e is dimensionless, so is $e^{-\alpha x^2}$.

▶ *What are the dimensions of* dx?

To find the dimensions of dx, follow the advice of Silvanus Thompson [45, p. 1]: Read d as "a little bit of." Then dx is "a little bit of x." A little length is still a length, so dx is a length. In general, dx has the same dimensions as x. Equivalently, d—the inverse of $\int$—is dimensionless.

Assembling the pieces, the whole integral has dimensions of length:

$$\left[\int e^{-\alpha x^2}\, dx\right] = \underbrace{\left[e^{-\alpha x^2}\right]}_{1} \times \underbrace{[dx]}_{L} = L. \tag{1.16}$$

> **Problem 1.7 Don't integrals compute areas?**
> A common belief is that integration computes areas. Areas have dimensions of L^2. How then can the Gaussian integral have dimensions of L?

1.3.3 Making an $f(\alpha)$ with correct dimensions

The third and final step in this dimensional analysis is to construct an $f(\alpha)$ with the same dimensions as the integral. Because the dimensions of α are L^{-2}, the only way to turn α into a length is to form $\alpha^{-1/2}$. Therefore,

$$f(\alpha) \sim \alpha^{-1/2}. \tag{1.17}$$

This useful result, which lacks only a dimensionless factor, was obtained without any integration.

To determine the dimensionless constant, set $\alpha = 1$ and evaluate

$$f(1) = \int_{-\infty}^{\infty} e^{-x^2}\, dx. \tag{1.18}$$

This classic integral will be approximated in Section 2.1 and guessed to be $\sqrt{\pi}$. The two results $f(1) = \sqrt{\pi}$ and $f(\alpha) \sim \alpha^{-1/2}$ require that $f(\alpha) = \sqrt{\pi/\alpha}$, which yields

$$\int_{-\infty}^{\infty} e^{-\alpha x^2}\, dx = \sqrt{\frac{\pi}{\alpha}}. \tag{1.19}$$

We often memorize the dimensionless constant but forget the power of α. Do not do that. The α factor is usually much more important than the dimensionless constant. Conveniently, the α factor is what dimensional analysis can compute.

> **Problem 1.8   Change of variable**
> Rewind back to page 8 and pretend that you do not know $f(\alpha)$. Without doing dimensional analysis, show that $f(\alpha) \sim \alpha^{-1/2}$.
>
> **Problem 1.9   Easy case $\alpha = 1$**
> Setting $\alpha = 1$, which is an example of easy-cases reasoning (Chapter 2), violates the assumption that x is a length and α has dimensions of L^{-2}. Why is it okay to set $\alpha = 1$?
>
> **Problem 1.10   Integrating a difficult exponential**
> Use dimensional analysis to investigate $\int_0^\infty e^{-\alpha t^3}\, dt$.

1.4 Summary and further problems

Do not add apples to oranges: Every term in an equation or sum must have identical dimensions! This restriction is a powerful tool. It helps us to evaluate integrals without integrating and to predict the solutions of differential equations. Here are further problems to practice this tool.

> **Problem 1.11   Integrals using dimensions**
> Use dimensional analysis to find $\int_0^\infty e^{-ax}\, dx$ and $\int \frac{dx}{x^2 + a^2}$. A useful result is
> $$\int \frac{dx}{x^2 + 1} = \arctan x + C. \qquad (1.20)$$
>
> **Problem 1.12   Stefan–Boltzmann law**
> Blackbody radiation is an electromagnetic phenomenon, so the radiation intensity depends on the speed of light c. It is also a thermal phenomenon, so it depends on the thermal energy $k_B T$, where T is the object's temperature and k_B is Boltzmann's constant. And it is a quantum phenomenon, so it depends on Planck's constant $\hbar$. Thus the blackbody-radiation intensity I depends on c, $k_B T$, and $\hbar$. Use dimensional analysis to show that $I \propto T^4$ and to find the constant of proportionality σ. Then look up the missing dimensionless constant. (These results are used in Section 5.3.3.)
>
> **Problem 1.13   Arcsine integral**
> Use dimensional analysis to find $\int \sqrt{1 - 3x^2}\, dx$. A useful result is
> $$\int \sqrt{1 - x^2}\, dx = \frac{\arcsin x}{2} + \frac{x\sqrt{1 - x^2}}{2} + C, \qquad (1.21)$$

Problem 1.14 Related rates

Water is poured into a large inverted cone (with a 90° opening angle) at a rate $dV/dt = 10\,\text{m}^3\,\text{s}^{-1}$. When the water depth is $h = 5\,\text{m}$, estimate the rate at which the depth is increasing. Then use calculus to find the exact rate.

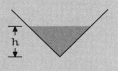

Problem 1.15 Kepler's third law

Newton's law of universal gravitation—the famous inverse-square law—says that the gravitational force between two masses is

$$F = -\frac{Gm_1 m_2}{r^2}, \qquad (1.22)$$

where G is Newton's constant, m_1 and m_2 are the two masses, and r is their separation. For a planet orbiting the sun, universal gravitation together with Newton's second law gives

$$m\frac{d^2\mathbf{r}}{dt^2} = -\frac{GMm}{r^2}\hat{\mathbf{r}}, \qquad (1.23)$$

where M is the mass of the sun, m the mass of the planet, $\mathbf{r}$ is the vector from the sun to the planet, and $\hat{\mathbf{r}}$ is the unit vector in the $\mathbf{r}$ direction.

How does the orbital period τ depend on orbital radius r? Look up Kepler's third law and compare your result to it.

2
Easy cases

2.1 Gaussian integral revisited	13
2.2 Plane geometry: The area of an ellipse	16
2.3 Solid geometry: The volume of a truncated pyramid	17
2.4 Fluid mechanics: Drag	21
2.5 Summary and further problems	29

A correct solution works in all cases, including the easy ones. This maxim underlies the second tool—the method of easy cases. It will help us guess integrals, deduce volumes, and solve exacting differential equations.

2.1 Gaussian integral revisited

As the first application, let's revisit the Gaussian integral from Section 1.3,

$$\int_{-\infty}^{\infty} e^{-\alpha x^2} \, dx. \tag{2.1}$$

▶ *Is the integral $\sqrt{\pi\alpha}$ or $\sqrt{\pi/\alpha}$?*

The correct choice must work for all $\alpha \geqslant 0$. At this range's endpoints ($\alpha = \infty$ and $\alpha = 0$), the integral is easy to evaluate.

▶ *What is the integral when $\alpha = \infty$?*

As the first easy case, increase α to ∞. Then $-\alpha x^2$ becomes very negative, even when x is tiny. The exponential of a large negative number is tiny, so the bell curve narrows to a sliver, and its area shrinks to zero. Therefore, as $\alpha \to \infty$ the integral shrinks to zero. This result refutes the option

$\sqrt{\pi\alpha}$, which is infinite when $\alpha = \infty$; and it supports the option $\sqrt{\pi/\alpha}$, which is zero when $\alpha = \infty$.

▶ *What is the integral when $\alpha = 0$?*

In the $\alpha = 0$ extreme, the bell curve flattens into a horizontal line with unit height. Its area, integrated over the infinite range, is infinite. This result refutes the $\sqrt{\pi\alpha}$ option, which is zero when $\alpha = 0$; and it supports the $\sqrt{\pi/\alpha}$ option, which is infinity when $\alpha = 0$. Thus the $\sqrt{\pi\alpha}$ option fails both easy-cases tests, and the $\sqrt{\pi/\alpha}$ option passes both easy-cases tests.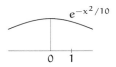

If these two options were the only options, we would choose $\sqrt{\pi/\alpha}$. However, if a third option were $\sqrt{2/\alpha}$, how could you decide between it and $\sqrt{\pi/\alpha}$? Both options pass both easy-cases tests; they also have identical dimensions. The choice looks difficult.

To choose, try a third easy case: $\alpha = 1$. Then the integral simplifies to

$$\int_{-\infty}^{\infty} e^{-x^2}\, dx. \tag{2.2}$$

This classic integral can be evaluated in closed form by using polar coordinates, but that method also requires a trick with few other applications (textbooks on multivariable calculus give the gory details). A less elegant but more general approach is to evaluate the integral numerically and to use the approximate value to guess the closed form.

Therefore, replace the smooth curve e^{-x^2} with a curve having n line segments. This piecewise-linear approximation turns the area into a sum of n trapezoids. As n approaches infinity, the area of the trapezoids more and more closely approaches the area under the smooth curve.

The table gives the area under the curve in the range $x = -10\ldots 10$, after dividing the curve into n line segments. The areas settle onto a stable value, and it looks familiar. It begins with 1.7, which might arise from $\sqrt{3}$. However, it continues as 1.77, which is too large to be $\sqrt{3}$. Fortunately, π is slightly larger than 3, so the area might be converging to $\sqrt{\pi}$.

n	Area
10	2.07326300569564
20	1.77263720482665
30	1.77245385170978
40	1.77245385090552
50	1.77245385090552

2.1 Gaussian integral revisited

Let's check by comparing the squared area against π:

$$1.77245385090552^2 \approx 3.14159265358980,$$
$$\pi \approx 3.14159265358979. \tag{2.3}$$

The close match suggests that the $\alpha = 1$ Gaussian integral is indeed $\sqrt{\pi}$:

$$\int_{-\infty}^{\infty} e^{-x^2}\, dx = \sqrt{\pi}. \tag{2.4}$$

Therefore the general Gaussian integral

$$\int_{-\infty}^{\infty} e^{-\alpha x^2}\, dx \tag{2.5}$$

must reduce to $\sqrt{\pi}$ when $\alpha = 1$. It must also behave correctly in the other two easy cases $\alpha = 0$ and $\alpha = \infty$.

Among the three choices $\sqrt{2/\alpha}$, $\sqrt{\pi/\alpha}$, and $\sqrt{\pi\alpha}$, only $\sqrt{\pi/\alpha}$ passes all three tests $\alpha = 0$, 1, and ∞. Therefore,

$$\int_{-\infty}^{\infty} e^{-\alpha x^2}\, dx = \sqrt{\frac{\pi}{\alpha}}. \tag{2.6}$$

Easy cases are not the only way to judge these choices. Dimensional analysis, for example, can also restrict the possibilities (Section 1.3). It even eliminates choices like $\sqrt{\pi/\alpha}$ that pass all three easy-cases tests. However, easy cases are, by design, simple. They do not require us to invent or deduce dimensions for x, α, and dx (the extensive analysis of Section 1.3). Easy cases, unlike dimensional analysis, can also eliminate choices like $\sqrt{2/\alpha}$ with correct dimensions. Each tool has its strengths.

> **Problem 2.1   Testing several alternatives**
> For the Gaussian integral
>
> $$\int_{-\infty}^{\infty} e^{-\alpha x^2}\, dx, \tag{2.7}$$
>
> use the three easy-cases tests to evaluate the following candidates for its value.
> (a) $\sqrt{\pi}/\alpha$ (b) $1 + (\sqrt{\pi} - 1)/\alpha$ (c) $1/\alpha^2 + (\sqrt{\pi} - 1)/\alpha$.
>
> **Problem 2.2   Plausible, incorrect alternative**
> Is there an alternative to $\sqrt{\pi/\alpha}$ that has valid dimensions and passes the three easy-cases tests?

> **Problem 2.3 Guessing a closed form**
> Use a change of variable to show that
> $$\int_0^\infty \frac{dx}{1+x^2} = 2\int_0^1 \frac{dx}{1+x^2}. \qquad (2.8)$$
> The second integral has a finite integration range, so it is easier than the first integral to evaluate numerically. Estimate the second integral using the trapezoid approximation and a computer or programmable calculator. Then guess a closed form for the first integral.

2.2 Plane geometry: The area of an ellipse

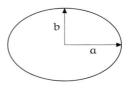

The second application of easy cases is from plane geometry: the area of an ellipse. This ellipse has semimajor axis a and semiminor axis b. For its area A consider the following candidates:

(a) ab^2 (b) $a^2 + b^2$ (c) a^3/b (d) $2ab$ (e) πab.

▶ *What are the merits or drawbacks of each candidate?*

The candidate $A = ab^2$ has dimensions of L^3, whereas an area must have dimensions of L^2. Thus ab^2 must be wrong.

The candidate $A = a^2 + b^2$ has correct dimensions (as do the remaining candidates), so the next tests are the easy cases of the radii a and b. For a, the low extreme $a = 0$ produces an infinitesimally thin ellipse with zero area. However, when $a = 0$ the candidate $A = a^2 + b^2$ reduces to $A = b^2$ rather than to 0; so $a^2 + b^2$ fails the $a = 0$ test.

The candidate $A = a^3/b$ correctly predicts zero area when $a = 0$. Because $a = 0$ was a useful easy case, and the axis labels a and b are almost interchangeable, its symmetric counterpart $b = 0$ should also be a useful easy case. It too produces an infinitesimally thin ellipse with zero area; alas, the candidate a^3/b predicts an infinite area, so it fails the $b = 0$ test. Two candidates remain.

The candidate $A = 2ab$ shows promise. When $a = 0$ or $b = 0$, the actual and predicted areas are zero, so $A = 2ab$ passes both easy-cases tests. Further testing requires the third easy case: $a = b$. Then the ellipse becomes a circle with radius a and area πa^2. The candidate $2ab$, however, reduces to $A = 2a^2$, so it fails the $a = b$ test.

The candidate $A = \pi ab$ passes all three tests: $a = 0$, $b = 0$, and $a = b$. With each passing test, our confidence in the candidate increases; and πab is indeed the correct area (Problem 2.4).

> **Problem 2.4 Area by calculus**
> Use integration to show that $A = \pi ab$.
>
> **Problem 2.5 Inventing a passing candidate**
> Can you invent a second candidate for the area that has correct dimensions and passes the $a = 0$, $b = 0$, and $a = b$ tests?
>
> **Problem 2.6 Generalization**
> Guess the volume of an ellipsoid with principal radii a, b, and c.

2.3 Solid geometry: The volume of a truncated pyramid

The Gaussian-integral example (Section 2.1) and the ellipse-area example (Section 2.2) showed easy cases as a method of analysis: for checking whether formulas are correct. The next level of sophistication is to use easy cases as a method of synthesis: for constructing formulas.

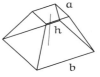

As an example, take a pyramid with a square base and slice a piece from its top using a knife parallel to the base. This truncated pyramid (called the frustum) has a square base and square top parallel to the base. Let h be its vertical height, b be the side length of its base, and a be the side length of its top.

▶ *What is the volume of the truncated pyramid?*

Let's synthesize the formula for the volume. It is a function of the three lengths h, a, and b. These lengths split into two kinds: height and base lengths. For example, flipping the solid on its head interchanges the meanings of a and b but preserves h; and no simple operation interchanges height with a or b. Thus the volume probably has two factors, each containing a length or lengths of only one kind:

$$V(h, a, b) = f(h) \times g(a, b). \tag{2.9}$$

Proportional reasoning will determine f; a bit of dimensional reasoning and a lot of easy-cases reasoning will determine g.

▶ *What is f: How should the volume depend on the height?*

To find f, use a proportional-reasoning thought experiment. Chop the solid into vertical slivers, each like an oil-drilling core; then imagine doubling h. This change doubles the volume of each sliver and therefore doubles the whole volume V. Thus $f \sim h$ and $V \propto h$: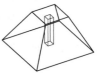

$$V = h \times g(a, b). \tag{2.10}$$

▶ *What is g: How should the volume depend on a and b?*

Because V has dimensions of L^3, the function $g(a, b)$ has dimensions of L^2. That constraint is all that dimensional analysis can say. Further constraints are needed to synthesize g, and these constraints are provided by the method of easy cases.

2.3.1 Easy cases

▶ *What are the easy cases of a and b?*

The easiest case is the extreme case $a = 0$ (an ordinary pyramid). The symmetry between a and b suggests two further easy cases, namely $a = b$ and the extreme case $b = 0$. The easy cases are then threefold:

$a = 0$

$b = 0$

$a = b$

When $a = 0$, the solid is an ordinary pyramid, and g is a function only of the base side length b. Because g has dimensions of L^2, the only possibility for g is $g \sim b^2$; in addition, $V \propto h$; so, $V \sim hb^2$. When $b = 0$, the solid is an upside-down version of the $b = 0$ pyramid and therefore has volume $V \sim ha^2$. When $a = b$, the solid is a rectangular prism having volume $V = ha^2$ (or hb^2).

▶ *Is there a volume formula that satisfies the three easy-cases constraints?*

2.3 Solid geometry: The volume of a truncated pyramid

The $a = 0$ and $b = 0$ constraints are satisfied by the symmetric sum $V \sim h(a^2 + b^2)$. If the missing dimensionless constant is $1/2$, making $V = h(a^2 + b^2)/2$, then the volume also satisfies the $a = b$ constraint, and the volume of an ordinary pyramid ($a = 0$) would be $hb^2/2$.

▶ *When $a = 0$, is the prediction $V = hb^2/2$ correct?*

Testing the prediction requires finding the exact dimensionless constant in $V \sim hb^2$. This task looks like a calculus problem: Slice a pyramid into thin horizontal sections and add (integrate) their volumes. However, a simple alternative is to apply easy cases again.

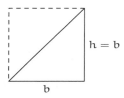

The easy case is easier to construct after we solve a similar but simpler problem: to find the area of a triangle with base b and height h. The area satisfies $A \sim hb$, but what is the dimensionless constant? To find it, choose b and h to make an easy triangle: a right triangle with $h = b$. Two such triangles make an easy shape: a square with area b^2. Thus each right triangle has area $A = b^2/2$; the dimensionless constant is $1/2$. Now extend this reasoning to three dimensions—find an ordinary pyramid (with a square base) that combines with itself to make an easy solid.

▶ *What is the easy solid?*

A convenient solid is suggested by the pyramid's square base: Perhaps each base is one face of a cube. The cube then requires six pyramids whose tips meet in the center of the cube; thus the pyramids have the aspect ratio $h = b/2$. For numerical simplicity, let's meet this condition with $b = 2$ and $h = 1$.

Six such pyramids form a cube with volume $b^3 = 8$, so the volume of one pyramid is $4/3$. Because each pyramid has volume $V \sim hb^2$, and $hb^2 = 4$ for these pyramids, the dimensionless constant in $V \sim hb^2$ must be $1/3$. The volume of an ordinary pyramid (a pyramid with $a = 0$) is therefore $V = hb^2/3$.

> **Problem 2.7 Triangular base**
> Guess the volume of a pyramid with height h and a triangular base of area A. Assume that the top vertex lies directly over the centroid of the base. Then try Problem 2.8.

Problem 2.8 Vertex location
The six pyramids do not make a cube unless each pyramid's top vertex lies directly above the center of the base. Thus the result $V = hb^2/3$ might apply only with this restriction. If instead the top vertex lies above one of the base vertices, what is the volume?

The prediction from the first three easy-cases tests was $V = hb^2/2$ (when $a = 0$), whereas the further easy case $h = b/2$ alongside $a = 0$ just showed that $V = hb^2/3$. The two methods are making contradictory predictions.

▶ *How can this contradiction be resolved?*

The contradiction must have snuck in during one of the reasoning steps. To find the culprit, revisit each step in turn. The argument for $V \propto h$ looks correct. The three easy-case requirements—that $V \sim hb^2$ when $a = 0$, that $V \sim ha^2$ when $b = 0$, and that $V = h(a^2 + b^2)/2$ when $a = b$—also look correct. The mistake was leaping from these constraints to the prediction $V \sim h(a^2 + b^2)$ for any a or b.

Instead let's try the following general form that includes an ab term:

$$V = h(\alpha a^2 + \beta ab + \gamma b^2). \tag{2.11}$$

Then solve for the coefficients α, β, and γ by reapplying the easy-cases requirements.

The $b = 0$ test along with the $h = b/2$ easy case, which showed that $V = hb^2/3$ for an ordinary pyramid, require that $\alpha = 1/3$. The $a = 0$ test similarly requires that $\gamma = 1/3$. And the $a = b$ test requires that $\alpha + \beta + \gamma = 1$. Therefore $\beta = 1/3$ and voilà,

$$V = \frac{1}{3}h(a^2 + ab + b^2). \tag{2.12}$$

This formula, the result of proportional reasoning, dimensional analysis, and the method of easy cases, is exact (Problem 2.9)!

Problem 2.9 Integration
Use integration to show that $V = h(a^2 + ab + b^2)/3$.

Problem 2.10 Truncated triangular pyramid
Instead of a pyramid with a square base, start with a pyramid with an equilateral triangle of side length b as its base. Then make the truncated solid by slicing a piece from the top using a knife parallel to the base. In terms of the height h

and the top and bottom side lengths a and b, what is the volume of this solid? (See also Problem 2.7.)

Problem 2.11   Truncated cone
What is the volume of a truncated cone with a circular base of radius r_1 and circular top of radius r_2 (with the top parallel to the base)? Generalize your formula to the volume of a truncated pyramid with height h, a base of an arbitrary shape and area A_{base}, and a corresponding top of area A_{top}.

2.4 Fluid mechanics: Drag

The preceding examples showed that easy cases can check and construct formulas, but the examples can be done without easy cases (for example, with calculus). For the next equations, from fluid mechanics, no exact solutions are known in general, so easy cases and other street-fighting tools are almost the only way to make progress.

Here then are the Navier–Stokes equations of fluid mechanics:

$$\frac{\partial \mathbf{v}}{\partial t} + (\mathbf{v} \cdot \nabla)\mathbf{v} = -\frac{1}{\rho}\nabla p + \nu \nabla^2 \mathbf{v}, \qquad (2.13)$$

where **v** is the velocity of the fluid (as a function of position and time), ρ is its density, p is the pressure, and ν is the kinematic viscosity. These equations describe an amazing variety of phenomena including flight, tornadoes, and river rapids.

Our example is the following home experiment on drag. Photocopy this page while magnifying it by a factor of 2; then cut out the following two templates:

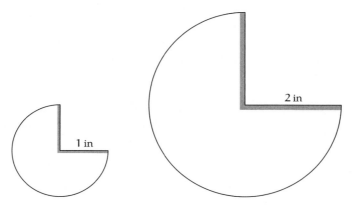

With each template, tape together the shaded areas to make a cone. The two resulting cones have the same shape, but the large cone has twice the height and width of the small cone.

▶ *When the cones are dropped point downward, what is the approximate ratio of their terminal speeds (the speeds at which drag balances weight)?*

The Navier–Stokes equations contain the answer to this question. Finding the terminal speed involves four steps.

Step 1. Impose boundary conditions. The conditions include the motion of the cone and the requirement that no fluid enters the paper.

Step 2. Solve the equations, together with the continuity equation $\nabla \cdot \mathbf{v} = 0$, in order to find the pressure and velocity at the surface of the cone.

Step 3. Use the pressure and velocity to find the pressure and velocity gradient at the surface of the cone; then integrate the resulting forces to find the net force and torque on the cone.

Step 4. Use the net force and torque to find the motion of the cone. This step is difficult because the resulting motion must be consistent with the motion assumed in step 1. If it is not consistent, go back to step 1, assume a different motion, and hope for better luck upon reaching this step.

Unfortunately, the Navier–Stokes equations are coupled and nonlinear partial-differential equations. Their solutions are known only in very simple cases: for example, a sphere moving very slowly in a viscous fluid, or a sphere moving at any speed in a zero-viscosity fluid. There is little hope of solving for the complicated flow around an irregular, quivering shape such as a flexible paper cone.

> **Problem 2.12   Checking dimensions in the Navier–Stokes equations**
> Check that the first three terms of the Navier–Stokes equations have identical dimensions.
>
> **Problem 2.13   Dimensions of kinematic viscosity**
> From the Navier–Stokes equations, find the dimensions of kinematic viscosity ν.

2.4 Fluid mechanics: Drag

2.4.1 Using dimensions

Because a direct solution of the Navier–Stokes equations is out of the question, let's use the methods of dimensional analysis and easy cases. A direct approach is to use them to deduce the terminal velocity itself. An indirect approach is to deduce the drag force as a function of fall speed and then to find the speed at which the drag balances the weight of the cones. This two-step approach simplifies the problem. It introduces only one new quantity (the drag force) but eliminates two quantities: the gravitational acceleration and the mass of the cone.

> **Problem 2.14 Explaining the simplification**
> Why is the drag force independent of the gravitational acceleration g and of the cone's mass m (yet the force depends on the cone's shape and size)?

The principle of dimensions is that all terms in a valid equation have identical dimensions. Applied to the drag force F, it means that in the equation $F = f(\text{quantities that affect } F)$ both sides have dimensions of force. Therefore, the strategy is to find the quantities that affect F, find their dimensions, and then combine the quantities into a quantity with dimensions of force.

▶ *On what quantities does the drag depend, and what are their dimensions?*

The drag force depends on four quantities: two parameters of the cone and two parameters of the fluid (air). (For the dimensions of ν, see Problem 2.13.)

v	speed of the cone	LT^{-1}
r	size of the cone	L
ρ	density of air	ML^{-3}
ν	viscosity of air	L^2T^{-1}

▶ *Do any combinations of the four parameters v, r, ρ, and ν have dimensions of force?*

The next step is to combine v, r, ρ, and ν into a quantity with dimensions of force. Unfortunately, the possibilities are numerous—for example,

$$F_1 = \rho v^2 r^2,$$
$$F_2 = \rho \nu v r, \qquad (2.14)$$

or the product combinations $\sqrt{F_1 F_2}$ and F_1^2/F_2. Any sum of these ugly products is also a force, so the drag force F could be $\sqrt{F_1 F_2} + F_1^2/F_2$, $3\sqrt{F_1 F_2} - 2F_1^2/F_2$, or much worse.

Narrowing the possibilities requires a method more sophisticated than simply guessing combinations with correct dimensions. To develop the sophisticated approach, return to the first principle of dimensions: All terms in an equation have identical dimensions. This principle applies to any statement about drag such as

$$A + B = C \tag{2.15}$$

where the blobs A, B, and C are functions of F, v, r, ρ, and ν.

Although the blobs can be absurdly complex functions, they have identical dimensions. Therefore, dividing each term by A, which produces the equation

$$\frac{A}{A} + \frac{B}{A} = \frac{C}{A}, \tag{2.16}$$

makes each term dimensionless. The same method turns any valid equation into a dimensionless equation. Thus, any (true) equation describing the world can be written in a dimensionless form.

Any dimensionless form can be built from dimensionless groups: from dimensionless products of the variables. Because any equation describing the world can be written in a dimensionless form, and any dimensionless form can be written using dimensionless groups, any equation describing the world can be written using dimensionless groups.

▶ *Is the free-fall example (Section 1.2) consistent with this principle?*

Before applying this principle to the complicated problem of drag, try it in the simple example of free fall (Section 1.2). The exact impact speed of an object dropped from a height h is $v = \sqrt{2gh}$, where g is the gravitational acceleration. This result can indeed be written in the dimensionless form $v/\sqrt{gh} = \sqrt{2}$, which itself uses only the dimensionless group $v/\sqrt{gh}$. The new principle passes its first test.

This dimensionless-group analysis of formulas, when reversed, becomes a method of synthesis. Let's warm up by synthesizing the impact speed v. First, list the quantities in the problem; here, they are v, g, and h. Second, combine these quantities into dimensionless groups. Here, all dimensionless groups can be constructed just from one group. For that group, let's choose v^2/gh (the particular choice does not affect the conclusion). Then the only possible dimensionless statement is

2.4 Fluid mechanics: Drag

$$\frac{v^2}{gh} = \text{dimensionless constant.} \tag{2.17}$$

(The right side is a dimensionless constant because no second group is available to use there.) In other words, $v^2/gh \sim 1$ or $v \sim \sqrt{gh}$.

This result reproduces the result of the less sophisticated dimensional analysis in Section 1.2. Indeed, with only one dimensionless group, either analysis leads to the same conclusion. However, in hard problems—for example, finding the drag force—the less sophisticated method does not provide its constraint in a useful form; then the method of dimensionless groups is essential.

> **Problem 2.15 Fall time**
> Synthesize an approximate formula for the free-fall time t from g and h.
>
> **Problem 2.16 Kepler's third law**
> Synthesize Kepler's third law connecting the orbital period of a planet to its orbital radius. (See also Problem 1.15.)

▶ *What dimensionless groups can be constructed for the drag problem?*

One dimensionless group could be $F/\rho v^2 r^2$; a second group could be rv/ν. Any other group can be constructed from these groups (Problem 2.17), so the problem is described by two *independent* dimensionless groups. The most general dimensionless statement is then

$$\text{one group} = f(\text{second group}), \tag{2.18}$$

where f is a still-unknown (but dimensionless) function.

▶ *Which dimensionless group belongs on the left side?*

The goal is to synthesize a formula for F, and F appears only in the first group $F/\rho v^2 r^2$. With that constraint in mind, place the first group on the left side rather than wrapping it in the still-mysterious function f. With this choice, the most general statement about drag force is

$$\frac{F}{\rho v^2 r^2} = f\left(\frac{rv}{\nu}\right). \tag{2.19}$$

The physics of the (steady-state) drag force on the cone is all contained in the dimensionless function f.

> **Problem 2.17 Only two groups**
> Show that F, v, r, ρ, and ν produce only two independent dimensionless groups.
>
> **Problem 2.18 How many groups in general?**
> Is there a general method to predict the number of independent dimensionless groups? (The answer was given in 1914 by Buckingham [9].)

The procedure might seem pointless, having produced a drag force that depends on the unknown function f. But it has greatly improved our chances of finding f. The original problem formulation required guessing the four-variable function h in $F = h(v, r, \rho, \nu)$, whereas dimensional analysis reduced the problem to guessing a function of only one variable (the ratio vr/ν). The value of this simplification was eloquently described by the statistician and physicist Harold Jeffreys (quoted in [34, p. 82]):

> A good table of functions of one variable may require a page; that of a function of two variables a volume; that of a function of three variables a bookcase; and that of a function of four variables a library.

> **Problem 2.19 Dimensionless groups for the truncated pyramid**
> The truncated pyramid of Section 2.3 has volume
>
> $$V = \frac{1}{3}h(a^2 + ab + b^2). \tag{2.20}$$
>
> Make dimensionless groups from V, h, a, and b, and rewrite the volume using these groups. (There are many ways to do so.)

2.4.2 Using easy cases

Although improved, our chances do not look high: Even the one-variable drag problem has no exact solution. But it might have exact solutions in its easy cases. Because the easiest cases are often extreme cases, look first at the extreme cases.

▶ *Extreme cases of what?*

The unknown function f depends on only rv/ν,

$$\frac{F}{\rho v^2 r^2} = f\left(\frac{rv}{\nu}\right), \tag{2.21}$$

so try extremes of rv/ν. However, to avoid lapsing into mindless symbol pushing, first determine the meaning of rv/ν. This combination rv/ν,

2.4 Fluid mechanics: Drag

often denoted Re, is the famous Reynolds number. (Its physical interpretation requires the technique of lumping and is explained in Section 3.4.3.)

The Reynolds number affects the drag force via the unknown function f:

$$\frac{F}{\rho v^2 r^2} = f(\text{Re}). \tag{2.22}$$

With luck, f can be deduced at extremes of the Reynolds number; with further luck, the falling cones are an example of one extreme.

▶ *Are the falling cones an extreme of the Reynolds number?*

The Reynolds number depends on r, v, and ν. For the speed v, everyday experience suggests that the cones fall at roughly $1\,\text{m s}^{-1}$ (within, say, a factor of 2). The size r is roughly $0.1\,\text{m}$ (again within a factor of 2). And the kinematic viscosity of air is $\nu \sim 10^{-5}\,\text{m}^2\,\text{s}^{-1}$. The Reynolds number is

$$\frac{\overbrace{0.1\,\text{m}}^{r} \times \overbrace{1\,\text{m s}^{-1}}^{v}}{\underbrace{10^{-5}\,\text{m}^2\,\text{s}^{-1}}_{\nu}} \sim 10^4. \tag{2.23}$$

It is significantly greater than 1, so the falling cones are an extreme case of high Reynolds number. (For low Reynolds number, try Problem 2.27 and see [38].)

> **Problem 2.20 Reynolds numbers in everyday flows**
> Estimate Re for a submarine cruising underwater, a falling pollen grain, a falling raindrop, and a 747 crossing the Atlantic.

The high-Reynolds-number limit can be reached many ways. One way is to shrink the viscosity ν to 0, because ν lives in the denominator of the Reynolds number. Therefore, in the limit of high Reynolds number, viscosity disappears from the problem and the drag force should not depend on viscosity. This reasoning contains several subtle untruths, yet its conclusion is mostly correct. (Clarifying the subtleties required two centuries of progress in mathematics, culminating in singular perturbations and the theory of boundary layers [12, 46].)

Viscosity affects the drag force only through the Reynolds number:

$$\frac{F}{\rho v^2 r^2} = f\left(\frac{rv}{\nu}\right). \tag{2.24}$$

To make F independent of viscosity, F must be independent of Reynolds number! The problem then contains only one independent dimensionless group, $F/\rho v^2 r^2$, so the most general statement about drag is

$$\frac{F}{\rho v^2 r^2} = \text{dimensionless constant.} \tag{2.25}$$

The drag force itself is then $F \sim \rho v^2 r^2$. Because r^2 is proportional to the cone's cross-sectional area A, the drag force is commonly written

$$F \sim \rho v^2 A. \tag{2.26}$$

Although the derivation was for falling cones, the result applies to any object as long as the Reynolds number is high. The shape affects only the missing dimensionless constant. For a sphere, it is roughly 1/4; for a long cylinder moving perpendicular to its axis, it is roughly 1/2; and for a flat plate moving perpendicular to its face, it is roughly 1.

2.4.3 Terminal velocities

The result $F \sim \rho v^2 A$ is enough to predict the terminal velocities of the cones. Terminal velocity means zero acceleration, so the drag force must balance the weight. The weight is $W = \sigma_{\text{paper}} A_{\text{paper}} g$, where σ_{paper} is the areal density of paper (mass per area) and A_{paper} is the area of the template after cutting out the quarter section. Because A_{paper} is comparable to the cross-sectional area A, the weight is roughly given by

$$W \sim \sigma_{\text{paper}} A g. \tag{2.27}$$

Therefore,

$$\underbrace{\rho v^2 A}_{\text{drag}} \sim \underbrace{\sigma_{\text{paper}} A g}_{\text{weight}}. \tag{2.28}$$

The area divides out and the terminal velocity becomes

$$v \sim \sqrt{\frac{g \sigma_{\text{paper}}}{\rho}}. \tag{2.29}$$

All cones constructed from the same paper and having the same shape, *whatever their size,* fall at the same speed!

To test this prediction, I constructed the small and large cones described on page 21, held one in each hand above my head, and let them fall. Their 2 m fall lasted roughly 2 s, and they landed within 0.1 s of one another. Cheap experiment and cheap theory agree!

> **Problem 2.21   Home experiment of a small versus a large cone**
> Try the cone home experiment yourself (page 21).
>
> **Problem 2.22   Home experiment of four stacked cones versus one cone**
> Predict the ratio
> $$\frac{\text{terminal velocity of four small cones stacked inside each other}}{\text{terminal velocity of one small cone}}. \quad (2.30)$$
> Test your prediction. Can you find a method not requiring timing equipment?
>
> **Problem 2.23   Estimating the terminal speed**
> Estimate or look up the areal density of paper; predict the cones' terminal speed; and then compare that prediction to the result of the home experiment.

2.5 Summary and further problems

A correct solution works in all cases, including the easy ones. Therefore, check any proposed formula in the easy cases, and guess formulas by constructing expressions that pass all easy-cases tests. To apply and extend these ideas, try the following problems and see the concise and instructive book by Cipra [10].

> **Problem 2.24   Fencepost errors**
> A garden has 10 m of horizontal fencing that you would like to divide into 1 m segments by using vertical posts. Do you need 10 or 11 vertical posts (including the posts needed at the ends)?
>
> **Problem 2.25   Odd sum**
> Here is the sum of the first n odd integers:
> $$S_n = \underbrace{1 + 3 + 5 + \cdots + l_n}_{n \text{ terms}} \quad (2.31)$$
>
> a. Does the last term l_n equal $2n + 1$ or $2n - 1$?
>
> b. Use easy cases to guess S_n (as a function of n).
>
> An alternative solution is discussed in Section 4.1.

Problem 2.26 Free fall with initial velocity

The ball in Section 1.2 was released from rest. Now imagine that it is given an initial velocity v_0 (where positive v_0 means an upward throw). Guess the impact velocity v_i.

Then solve the free-fall differential equation to find the exact v_i, and compare the exact result to your guess.

Problem 2.27 Low Reynolds number

In the limit $\mathrm{Re} \ll 1$, guess the form of f in

$$\frac{F}{\rho v^2 r^2} = f\left(\frac{rv}{\nu}\right). \tag{2.32}$$

The result, when combined with the correct dimensionless constant, is known as Stokes drag [12].

Problem 2.28 Range formula

How far does a rock travel horizontally (no air resistance)? Use dimensions and easy cases to guess a formula for the range R as a function of the launch velocity v, the launch angle θ, and the gravitational acceleration g.

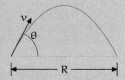

Problem 2.29 Spring equation

The angular frequency of an ideal mass–spring system (Section 3.4.2) is $\sqrt{k/m}$, where k is the spring constant and m is the mass. This expression has the spring constant k in the numerator. Use extreme cases of k or m to decide whether that placement is correct.

Problem 2.30 Taping the cone templates

The tape mark on the large cone template (page 21) is twice as wide as the tape mark on the small cone template. In other words, if the tape on the large cone is, say, 6 mm wide, the tape on the small cone should be 3 mm wide. Why?

3
Lumping

3.1	Estimating populations: How many babies?	32
3.2	Estimating integrals	33
3.3	Estimating derivatives	37
3.4	Analyzing differential equations: The spring–mass system	42
3.5	Predicting the period of a pendulum	46
3.6	*Summary and further problems*	54

Where will an orbiting planet be 6 months from now? To predict its new location, we cannot simply multiply the 6 months by the planet's current velocity, for its velocity constantly varies. Such calculations are the reason that calculus was invented. Its fundamental idea is to divide the time into tiny intervals over which the velocity is constant, to multiply each tiny time by the corresponding velocity to compute a tiny distance, and then to add the tiny distances.

Amazingly, this computation can often be done exactly, even when the intervals have infinitesimal width and are therefore infinite in number. However, the symbolic manipulations can be lengthy and, worse, are often rendered impossible by even small changes to the problem. Using calculus methods, for example, we can exactly calculate the area under the Gaussian e^{-x^2} between $x = 0$ and ∞; yet if either limit is any value except zero or infinity, an exact calculation becomes impossible.

In contrast, approximate methods are robust: They almost always provide a reasonable answer. And the least accurate but most robust method is lumping. Instead of dividing a changing process into many tiny pieces, group or lump it into one or two pieces. This simple approximation and its advantages are illustrated using examples ranging from demographics (Section 3.1) to nonlinear differential equations (Section 3.5).

3.1 Estimating populations: How many babies?

The first example is to estimate the number of babies in the United States. For definiteness, call a child a baby until he or she turns 2 years old. An exact calculation requires the birth dates of every person in the United States. This, or closely similar, information is collected once every decade by the US Census Bureau.

As an approximation to this voluminous data, the Census Bureau [47] publishes the number of people at each age. The data for 1991 is a set of points lying on a wiggly line N(t), where t is age. Then

$$N_{babies} = \int_0^{2\,yr} N(t)\,dt. \qquad (3.1)$$

Problem 3.1 Dimensions of the vertical axis
Why is the vertical axis labeled in units of people per year rather than in units of people? Equivalently, why does the axis have dimensions of T^{-1}?

This method has several problems. First, it depends on the huge resources of the US Census Bureau, so it is not usable on a desert island for back-of-the-envelope calculations. Second, it requires integrating a curve with no analytic form, so the integration must be done numerically. Third, the integral is of data specific to this problem, whereas mathematics should be about generality. An exact integration, in short, provides little insight and has minimal transfer value. Instead of integrating the population curve exactly, approximate it—lump the curve into one rectangle.

▶ *What are the height and width of this rectangle?*

The rectangle's width is a time, and a plausible time related to populations is the life expectancy. It is roughly 80 years, so make 80 years the width by pretending that everyone dies abruptly on his or her 80th birthday. The rectangle's height can be computed from the rectangle's area, which is the US population—conveniently 300 million in 2008. Therefore,

$$\text{height} = \frac{\text{area}}{\text{width}} \sim \frac{3 \times 10^8}{75\,\text{yr}}. \qquad (3.2)$$

▶ *Why did the life expectancy drop from 80 to 75 years?*

Fudging the life expectancy simplifies the mental division: 75 divides easily into 3 and 300. The inaccuracy is no larger than the error made by lumping, and it might even cancel the lumping error. Using 75 years as the width makes the height approximately $4 \times 10^6 \text{ yr}^{-1}$.

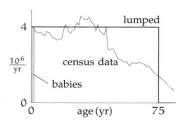

Integrating the population curve over the range $t = 0 \ldots 2 \text{ yr}$ becomes just multiplication:

$$N_{babies} \sim \underbrace{4 \times 10^6 \text{ yr}^{-1}}_{\text{height}} \times \underbrace{2 \text{ yr}}_{\text{infancy}} = 8 \times 10^6. \tag{3.3}$$

The Census Bureau's figure is very close: 7.980×10^6. The error from lumping canceled the error from fudging the life expectancy to 75 years!

> **Problem 3.2 Landfill volume**
> Estimate the US landfill volume used annually by disposable diapers.
>
> **Problem 3.3 Industry revenues**
> Estimate the annual revenue of the US diaper industry.

3.2 Estimating integrals

The US population curve (Section 3.1) was difficult to integrate partly because it was unknown. But even well-known functions can be difficult to integrate. In such cases, two lumping methods are particularly useful: the $1/e$ heuristic (Section 3.2.1) and the full width at half maximum (FWHM) heuristic (Section 3.2.2).

3.2.1 $1/e$ heuristic

Electronic circuits, atmospheric pressure, and radioactive decay contain the ubiquitous exponential and its integral (given here in dimensionless form)

$$\int_0^\infty e^{-t} \, dt. \tag{3.4}$$

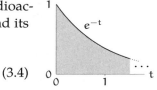

To approximate its value, let's lump the e^{-t} curve into one rectangle.

▶ *What values should be chosen for the width and height of the rectangle?*

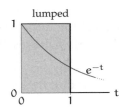

A reasonable height for the rectangle is the maximum of e^{-t}, namely 1. To choose its width, use significant change as the criterion (a method used again in Section 3.3.3): Choose a significant change in e^{-t}; then find the width Δt that produces this change. In an exponential decay, a simple and natural significant change is when e^{-t} becomes a factor of e closer to its final value (which is 0 here because t goes to ∞). With this criterion, $\Delta t = 1$. The lumping rectangle then has unit area—which is the exact value of the integral!

Encouraged by this result, let's try the heuristic on the difficult integral

$$\int_{-\infty}^{\infty} e^{-x^2} \, dx. \tag{3.5}$$

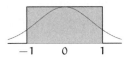

Again lump the area into a single rectangle. Its height is the maximum of e^{-x^2}, which is 1. Its width is enough that e^{-x^2} falls by a factor of e. This drop happens at $x = \pm 1$, so the width is $\Delta x = 2$ and its area is 1×2. The exact area is $\sqrt{\pi} \approx 1.77$ (Section 2.1), so lumping makes an error of only 13%: For such a short derivation, the accuracy is extremely high.

Problem 3.4 General exponential decay
Use lumping to estimate the integral

$$\int_0^\infty e^{-\alpha t} \, dt. \tag{3.6}$$

Use dimensional analysis and easy cases to check that your answer makes sense.

Problem 3.5 Atmospheric pressure
Atmospheric density ρ decays roughly exponentially with height z:

$$\rho \sim \rho_0 e^{-z/H}, \tag{3.7}$$

where ρ_0 is the density at sea level, and H is the so-called scale height (the height at which the density falls by a factor of e). Use your everyday experience to estimate H.

Then estimate the atmospheric pressure at sea level by estimating the weight of an infinitely high cylinder of air.

Problem 3.6 Cone free-fall distance
Roughly how far does a cone of Section 2.4 fall before reaching a significant fraction of its terminal velocity? How large is that distance compared to the drop height of 2 m? *Hint:* Sketch (very roughly) the cone's acceleration versus time and make a lumping approximation.

3.2.2 Full width at half maximum

Another reasonable lumping heuristic arose in the early days of spectroscopy. As a spectroscope swept through a range of wavelengths, a chart recorder would plot how strongly a molecule absorbed radiation of that wavelength. This curve contains many peaks whose location and area reveal the structure of the molecule (and were essential in developing quantum theory [14]). But decades before digital chart recorders existed, how could the areas of the peaks be computed?

They were computed by lumping the peak into a rectangle whose height is the height of the peak and whose width is the full width at half maximum (FWHM). Where the $1/e$ heuristic uses a factor of e as the significant change, the FWHM heuristic uses a factor of 2.

Try this recipe on the Gaussian integral $\int_{-\infty}^{\infty} e^{-x^2}\, dx$.

The maximum height of e^{-x^2} is 1, so the half maxima are at $x = \pm\sqrt{\ln 2}$ and the full width is $2\sqrt{\ln 2}$. The lumped rectangle therefore has area $2\sqrt{\ln 2} \approx 1.665$. The exact area is $\sqrt{\pi} \approx 1.77$ (Section 2.1): The FWHM heuristic makes an error of only 6%, which is roughly one-half the error of the $1/e$ heuristic.

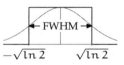

Problem 3.7 Trying the FWHM heuristic
Make single-rectangle lumping estimates of the following integrals. Choose the height and width of the rectangle using the FWHM heuristic. How accurate is each estimate?

a. $\displaystyle\int_{-\infty}^{\infty} \frac{1}{1+x^2}\, dx$ [exact value: π].

b. $\displaystyle\int_{-\infty}^{\infty} e^{-x^4}\, dx$ [exact value: $\Gamma(1/4)/2 \approx 1.813$].

3.2.3 Stirling's approximation

The 1/e and FWHM lumping heuristics next help us approximate the ubiquitous factorial function n!; this function's uses range from probability theory to statistical mechanics and the analysis of algorithms. For positive integers, n! is defined as $n \times (n-1) \times (n-2) \times \cdots \times 2 \times 1$. In this discrete form, it is difficult to approximate. However, the integral representation for n!,

$$n! \equiv \int_0^\infty t^n e^{-t}\, dt, \qquad (3.8)$$

provides a definition even when n is not a positive integer—and this integral can be approximated using lumping.

The lumping analysis will generate almost all of Stirling's famous approximation formula

$$n! \approx n^n e^{-n} \sqrt{2\pi n}. \qquad (3.9)$$

▶ *Lumping requires a peak, but does the integrand $t^n e^{-t}$ have a peak?*

To understand the integrand $t^n e^{-t}$ or t^n/e^t, examine the extreme cases of t. When $t = 0$, the integrand is 0. In the opposite extreme, $t \to \infty$, the polynomial factor t^n makes the product infinity while the exponential factor e^{-t} makes it zero. Who wins that struggle? The Taylor series for e^t contains every power of t (and with positive coefficients), so it is an increasing, infinite-degree polynomial. Therefore, as t goes to infinity, e^t outruns any polynomial t^n and makes the integrand t^n/e^t equal 0 in the $t \to \infty$ extreme. Being zero at both extremes, the integrand must have a peak in between. In fact, it has exactly one peak. (Can you show that?)

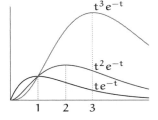

Increasing n strengthens the polynomial factor t^n, so t^n survives until higher t before e^t outruns it. Therefore, the peak of t^n/e^t shifts right as n increases. The graph confirms this prediction and suggests that the peak occurs at $t = n$. Let's check by using calculus to maximize $t^n e^{-t}$ or, more simply, to maximize its logarithm $f(t) = n \ln t - t$. At a peak, a function has zero slope. Because $df/dt = n/t - 1$, the peak occurs at $t_{\text{peak}} = n$, when the integrand $t^n e^{-t}$ is $n^n e^{-n}$—thus reproducing the largest and most important factor in Stirling's formula.

3.3 Estimating derivatives

▶ *What is a reasonable lumping rectangle?*

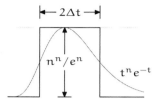

The rectangle's height is the peak height $n^n e^{-n}$. For the rectangle's width, use either the $1/e$ or the FWHM heuristics. Because both heuristic require approximating $t^n e^{-t}$, expand its logarithm $f(t)$ in a Taylor series around its peak at $t = n$:

$$f(n + \Delta t) = f(n) + \Delta t \frac{df}{dt}\bigg|_{t=n} + \frac{(\Delta t)^2}{2} \frac{d^2 f}{dt^2}\bigg|_{t=n} + \cdots. \qquad (3.10)$$

The second term of the Taylor expansion vanishes because $f(t)$ has zero slope at the peak. In the third term, the second derivative $d^2 f/dt^2$ at $t = n$ is $-n/t^2$ or $-1/n$. Thus,

$$f(n + \Delta t) \approx f(n) - \frac{(\Delta t)^2}{2n}. \qquad (3.11)$$

To decrease $t^n e^{-t}$ by a factor of F requires decreasing $f(t)$ by $\ln F$. This choice means $\Delta t = \sqrt{2n \ln F}$. Because the rectangle's width is $2\Delta t$, the lumped-area estimate of $n!$ is

$$n! \sim n^n e^{-n} \sqrt{n} \times \begin{cases} \sqrt{8} & (1/e \text{ criterion: } F = e) \\ \sqrt{8 \ln 2} & (\text{FWHM criterion: } F = 2). \end{cases} \qquad (3.12)$$

For comparison, Stirling's formula is $n! \approx n^n e^{-n} \sqrt{2\pi n}$. Lumping has explained almost every factor. The $n^n e^{-n}$ factor is the height of the rectangle, and the $\sqrt{n}$ factor is from the width of the rectangle. Although the exact $\sqrt{2\pi}$ factor remains mysterious (Problem 3.9), it is approximated to within 13% (the $1/e$ heuristic) or 6% (the FWHM heuristic).

> **Problem 3.8 Coincidence?**
> The FWHM approximation for the area under a Gaussian (Section 3.2.2) was also accurate to 6%. Coincidence?
>
> **Problem 3.9 Exact constant in Stirling's formula**
> Where does the more accurate constant factor of $\sqrt{2\pi}$ come from?

3.3 Estimating derivatives

In the preceding examples, lumping helped estimate integrals. Because integration and differentiation are closely related, lumping also provides a method for estimating derivatives. The method begins with a dimensional observation about derivatives. A derivative is a ratio of differentials; for example, df/dx is the ratio of df to dx. Because d is dimensionless (Section 1.3.2), the dimensions of df/dx are the dimensions of f/x. This useful, surprising conclusion is worth testing with a familiar example: Differentiating height y with respect to time t produces velocity dy/dt, whose dimensions of LT^{-1} are indeed the dimensions of y/t.

> **Problem 3.10 Dimensions of a second derivative**
> What are the dimensions of d^2f/dx^2?

3.3.1 Secant approximation

As df/dx and f/x have identical dimensions, perhaps their magnitudes are similar:

$$\frac{df}{dx} \sim \frac{f}{x}. \qquad (3.13)$$

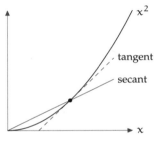

Geometrically, the derivative df/dx is the slope of the tangent line, whereas the approximation f/x is the slope of the secant line. By replacing the curve with the secant line, we make a lumping approximation.

Let's test the approximation on an easy function such as $f(x) = x^2$. Good news—the secant and tangent slopes differ only by a factor of 2:

$$\frac{df}{dx} = 2x \quad \text{and} \quad \frac{f(x)}{x} = x. \qquad (3.14)$$

> **Problem 3.11 Higher powers**
> Investigate the secant approximation for $f(x) = x^n$.
>
> **Problem 3.12 Second derivatives**
> Use the secant approximation to estimate d^2f/dx^2 with $f(x) = x^2$. How does the approximation compare to the exact second derivative?

▶ *How accurate is the secant approximation for $f(x) = x^2 + 100$?*

The secant approximation is quick and useful but can make large errors. When $f(x) = x^2 + 100$, for example, the secant and tangent at $x = 1$

3.3 Estimating derivatives

have dramatically different slopes. The tangent slope df/dx is 2, whereas the secant slope f(1)/1 is 101. The ratio of these two slopes, although dimensionless, is distressingly large.

> **Problem 3.13 Investigating the discrepancy**
>
> With $f(x) = x^2 + 100$, sketch the ratio
>
> $$\frac{\text{secant slope}}{\text{tangent slope}} \qquad (3.15)$$
>
> as a function of x. The ratio is not constant! Why is the dimensionless factor not constant? (That question is tricky.)

The large discrepancy in replacing the derivative df/dx, which is

$$\lim_{\Delta x \to 0} \frac{f(x) - f(x - \Delta x)}{\Delta x}, \qquad (3.16)$$

with the secant slope f(x)/x is due to two approximations. The first approximation is to take $\Delta x = x$ rather than $\Delta x = 0$. Then df/dx ≈ (f(x) − f(0))/x. This first approximation produces the slope of the line from (0, f(0)) to (x, f(x)). The second approximation replaces f(0) with 0, which produces df/dx ≈ f/x; that ratio is the slope of the secant from (0, 0) to (x, f(x)).

3.3.2 Improved secant approximation

The second approximation is fixed by starting the secant at (0, f(0)) instead of (0, 0).

▸ *With that change, what are the secant and tangent slopes when $f(x) = x^2 + C$?*

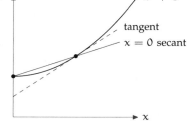

Call the secant starting at (0, 0) the origin secant; call the new secant the x = 0 secant. Then the x = 0 secant always has one-half the slope of the tangent, no matter the constant C. The x = 0 secant approximation is robust against—is unaffected by—vertical translation.

▸ *How robust is the x = 0 secant approximation against horizontal translation?*

To investigate how the x = 0 secant handles horizontal translation, translate $f(x) = x^2$ rightward by 100 to make $f(x) = (x-100)^2$. At the parabola's vertex $x = 100$, the $x = 0$ secant, from $(0, 10^4)$ to $(100, 0)$, has slope -100; however, the tangent has zero slope. Thus the $x = 0$ secant, although an improvement on the origin secant, is affected by horizontal translation.

3.3.3 Significant-change approximation

The derivative itself is unaffected by horizontal and vertical translation, so a derivative suitably approximated might be translation invariant. An approximate derivative is

$$\frac{df}{dx} \approx \frac{f(x + \Delta x) - f(x)}{\Delta x}, \tag{3.17}$$

where Δx is not zero but is still small.

▶ *How small should Δx be? Is $\Delta x = 0.01$ small enough?*

The choice $\Delta x = 0.01$ has two defects. First, it cannot work when x has dimensions. If x is a length, what length is small enough? Choosing $\Delta x = 1$ mm is probably small enough for computing derivatives related to the solar system, but is probably too large for computing derivatives related to falling fog droplets. Second, no fixed choice can be scale invariant. Although $\Delta x = 0.01$ produces accurate derivatives when $f(x) = \sin x$, it fails when $f(x) = \sin 1000x$, the result of simply rescaling x to $1000x$.

These problems suggest trying the following significant-change approximation:

$$\frac{df}{dx} \sim \frac{\text{significant } \Delta f \text{ (change in f) at } x}{\Delta x \text{ that produces a significant } \Delta f}. \tag{3.18}$$

Because the Δx here is defined by the properties of the curve at the point of interest, without favoring particular coordinate values or values of Δx, the approximation is scale and translation invariant.

To illustrate this approximation, let's try $f(x) = \cos x$ and estimate df/dx at $x = 3\pi/2$ with the three approximations: the origin secant, the $x = 0$ secant, and the significant-change approximation. The origin secant goes from $(0, 0)$ to $(3\pi/2, 0)$, so it has zero slope. It is a poor approximation to the exact slope of 1. The $x = 0$

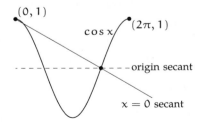

3.3 Estimating derivatives

secant goes from $(0,1)$ to $(3\pi/2, 0)$, so it has a slope of $-2/3\pi$, which is worse than predicting zero slope because even the sign is wrong!

The significant-change approximation might provide more accuracy. What is a significant change in $f(x) = \cos x$? Because the cosine changes by 2 (from -1 to 1), call $1/2$ a significant change in $f(x)$. That change happens when x changes from $3\pi/2$, where $f(x) = 0$, to $3\pi/2 + \pi/6$, where $f(x) = 1/2$. In other words, Δx is $\pi/6$. The approximate derivative is therefore

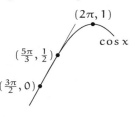

$$\frac{df}{dx} \sim \frac{\text{significant } \Delta f \text{ near } x}{\Delta x} \sim \frac{1/2}{\pi/6} = \frac{3}{\pi}. \tag{3.19}$$

This estimate is approximately 0.955—amazingly close to the true derivative of 1.

Problem 3.14 Derivative of a quadratic
With $f(x) = x^2$, estimate df/dx at $x = 5$ using three approximations: the origin secant, the $x = 0$ secant, and the significant-change approximation. Compare these estimates to the true slope.

Problem 3.15 Derivative of the logarithm
Use the significant-change approximation to estimate the derivative of $\ln x$ at $x = 10$. Compare the estimate to the true slope.

Problem 3.16 Lennard–Jones potential
The Lennard–Jones potential is a model of the interaction energy between two nonpolar molecules such as N_2 or CH_4. It has the form

$$V(r) = 4\epsilon \left[\left(\frac{\sigma}{r}\right)^{12} - \left(\frac{\sigma}{r}\right)^{6} \right], \tag{3.20}$$

where r is the distance between the molecules, and ϵ and σ are constants that depend on the molecules. Use the origin secant to estimate r_0, the separation r at which $V(r)$ is a minimum. Compare the estimate to the true r_0 found using calculus.

Problem 3.17 Approximate maxima and minima
Let $f(x)$ be an increasing function and $g(x)$ a decreasing function. Use the origin secant to show, approximately, that $h(x) = f(x) + g(x)$ has a minimum where $f(x) = g(x)$. This useful rule of thumb, which generalizes Problem 3.16, is often called the balancing heuristic.

3.4 Analyzing differential equations: The spring–mass system

Estimating derivatives reduces differentiation to division (Section 3.3); it thereby reduces differential equations to algebraic equations.

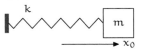

To produce an example equation to analyze, connect a block of mass m to an ideal spring with spring constant (stiffness) k, pull the block a distance x_0 to the right relative to the equilibrium position $x = 0$, and release it at time $t = 0$. The block oscillates back and forth, its position x described by the ideal-spring differential equation

$$m\frac{d^2x}{dt^2} + kx = 0. \tag{3.21}$$

Let's approximate the equation and thereby estimate the oscillation frequency.

3.4.1 Checking dimensions

Upon seeing any equation, first check its dimensions (Chapter 1). If all terms do not have identical dimensions, the equation is not worth solving—a great savings of effort. If the dimensions match, the check has prompted reflection on the meaning of the terms; this reflection helps prepare for solving the equation and for understanding any solution.

▶ *What are the dimensions of the two terms in the spring equation?*

Look first at the simple second term kx. It arises from Hooke's law, which says that an ideal spring exerts a force kx where x is the extension of the spring relative to its equilibrium length. Thus the second term kx is a force. Is the first term also a force?

The first term $m(d^2x/dt^2)$ contains the second derivative d^2x/dt^2, which is familiar as an acceleration. Many differential equations, however, contain unfamiliar derivatives. The Navier–Stokes equations of fluid mechanics (Section 2.4),

$$\frac{\partial \mathbf{v}}{\partial t} + (\mathbf{v}\cdot\nabla)\mathbf{v} = -\frac{1}{\rho}\nabla p + \nu\nabla^2\mathbf{v}, \tag{3.22}$$

contain two strange derivatives: $(\mathbf{v}\cdot\nabla)\mathbf{v}$ and $\nabla^2\mathbf{v}$. What are the dimensions of those terms?

3.4 Analyzing differential equations: The spring–mass system

To practice for later handling such complicated terms, let's now find the dimensions of d^2x/dt^2 by hand. Because d^2x/dt^2 contains two exponents of 2, and x is length and t is time, d^2x/dt^2 might plausibly have dimensions of L^2T^{-2}.

▶ *Are L^2T^{-2} the correct dimensions?*

To decide, use the idea from Section 1.3.2 that the differential symbol d means "a little bit of." The numerator d^2x, meaning d of dx, is "a little bit of a little bit of x." Thus, it is a length. The denominator dt^2 could plausibly mean $(dt)^2$ or $d(t^2)$. [It turns out to mean $(dt)^2$.] In either case, its dimensions are T^2. Therefore, the dimensions of the second derivative are LT^{-2}:

$$\left[\frac{d^2x}{dt^2}\right] = LT^{-2}. \tag{3.23}$$

This combination is an acceleration, so the spring equation's first term $m(d^2x/dt^2)$ is mass times acceleration—giving it the same dimensions as the kx term.

> **Problem 3.18** Dimensions of spring constant
> What are the dimensions of the spring constant k?

3.4.2 Estimating the magnitudes of the terms

The spring equation passes the dimensions test, so it is worth analyzing to find the oscillation frequency. The method is to replace each term with its approximate magnitude. These replacements will turn a complicated differential equation into a simple algebraic equation for the frequency.

To approximate the first term $m(d^2x/dt^2)$, use the significant-change approximation (Section 3.3.3) to estimate the magnitude of the acceleration d^2x/dt^2.

$$\frac{d^2x}{dt^2} \sim \frac{\text{significant } \Delta x}{(\Delta t \text{ that produces a significant } \Delta x)^2}. \tag{3.24}$$

> **Problem 3.19** Explaining the exponents
> The numerator contains only the first power of Δx, whereas the denominator contains the second power of Δt. How can that discrepancy be correct?

To evaluate this approximate acceleration, first decide on a significant Δx—on what constitutes a significant change in the mass's position. The mass moves between the points $x = -x_0$ and $x = +x_0$, so a significant change in position should be a significant fraction of the peak-to-peak amplitude $2x_0$. The simplest choice is $\Delta x = x_0$.

Now estimate Δt: the time for the block to move a distance comparable to Δx. This time—called the characteristic time of the system—is related to the oscillation period T. During one period, the mass moves back and forth and travels a distance $4x_0$—much farther than x_0. If Δt were, say, $T/4$ or $T/2\pi$, then in the time Δt the mass would travel a distance comparable to x_0. Those choices for Δt have a natural interpretation as being approximately $1/\omega$, where the angular frequency ω is connected to the period by the definition $\omega \equiv 2\pi/T$. With the preceding choices for Δx and Δt, the $m(d^2x/dt^2)$ term is roughly $mx_0\omega^2$.

▶ *What does "is roughly" mean?*

The phrase cannot mean that $mx_0\omega^2$ and $m(d^2x/dt^2)$ are within, say, a factor of 2, because $m(d^2x/dt^2)$ varies and mx_0/τ^2 is constant. Rather, "is roughly" means that a typical or characteristic magnitude of $m(d^2x/dt^2)$—for example, its root-mean-square value—is comparable to $mx_0\omega^2$. Let's include this meaning within the twiddle notation $\sim$. Then the typical-magnitude estimate can be written

$$m\frac{d^2x}{dt^2} \sim mx_0\omega^2. \tag{3.25}$$

With the same meaning of "is roughly", namely that the typical magnitudes are comparable, the spring equation's second term kx is roughly kx_0. The two terms must add to zero—a consequence of the spring equation

$$m\frac{d^2x}{dt^2} + kx = 0. \tag{3.26}$$

Therefore, the magnitudes of the two terms are comparable:

$$mx_0\omega^2 \sim kx_0. \tag{3.27}$$

The amplitude x_0 divides out! With x_0 gone, the frequency ω and oscillation period $T = 2\pi/\omega$ are independent of amplitude. [This reasoning uses several approximations, but this conclusion is exact (Problem 3.20).] The approximated angular frequency ω is then $\sqrt{k/m}$.

For comparison, the exact solution of the spring differential equation is, from Problem 3.22,

$$x = x_0 \cos \omega t, \qquad (3.28)$$

where ω is $\sqrt{k/m}$. The approximated angular frequency is also exact!

> **Problem 3.20 Amplitude independence**
> Use dimensional analysis to show that the angular frequency ω cannot depend on the amplitude x_0.
>
> **Problem 3.21 Checking dimensions in the alleged solution**
> What are the dimensions of ωt? What are the dimensions of $\cos \omega t$? Check the dimensions of the proposed solution $x = x_0 \cos \omega t$, and the dimensions of the proposed period $2\pi\sqrt{m/k}$.
>
> **Problem 3.22 Verification**
> Show that $x = x_0 \cos \omega t$ with $\omega = \sqrt{k/m}$ solves the spring differential equation
>
> $$m\frac{d^2x}{dt^2} + kx = 0. \qquad (3.29)$$

3.4.3 Meaning of the Reynolds number

As a further example of lumping—in particular, of the significant-change approximation—let's analyze the Navier–Stokes equations introduced in Section 2.4,

$$\frac{\partial \mathbf{v}}{\partial t} + (\mathbf{v}\cdot\nabla)\mathbf{v} = -\frac{1}{\rho}\nabla p + \nu\nabla^2\mathbf{v}, \qquad (3.30)$$

and extract from them a physical meaning for the Reynolds number rv/ν.

To do so, we estimate the typical magnitude of the inertial term $(\mathbf{v}\cdot\nabla)\mathbf{v}$ and of the viscous term $\nu\nabla^2\mathbf{v}$.

▶ *What is the typical magnitude of the inertial term?*

The inertial term $(\mathbf{v}\cdot\nabla)\mathbf{v}$ contains the spatial derivative $\nabla\mathbf{v}$. According to the significant-change approximation (Section 3.3.3), the derivative $\nabla\mathbf{v}$ is roughly the ratio

$$\frac{\text{significant change in flow velocity}}{\text{distance over which flow velocity changes significantly}}. \qquad (3.31)$$

The flow velocity (the velocity of the air) is nearly zero far from the cone and is comparable to v near the cone (which is moving at speed v). Therefore, v, or a reasonable fraction of v, constitutes a significant change in flow velocity. This speed change happens over a distance comparable to the size of the cone: Several cone lengths away, the air hardly knows about the falling cone. Thus $\nabla \mathbf{v} \sim v/r$. The inertial term $(\mathbf{v} \cdot \nabla)\mathbf{v}$ contains a second factor of $\mathbf{v}$, so $(\mathbf{v} \cdot \nabla)\mathbf{v}$ is roughly v^2/r.

▶ *What is the typical magnitude of the viscous term?*

The viscous term $\nu \nabla^2 \mathbf{v}$ contains two spatial derivatives of $\mathbf{v}$. Because each spatial derivative contributes a factor of $1/r$ to the typical magnitude, $\nu \nabla^2 \mathbf{v}$ is roughly $\nu v/r^2$. The ratio of the inertial term to the viscous term is then roughly $(v^2/r)/(\nu v/r^2)$. This ratio simplifies to rv/ν—the familiar, dimensionless, Reynolds number.

Thus, the Reynolds number measures the importance of viscosity. When Re $\gg$ 1, the viscous term is small, and viscosity has a negligible effect. It cannot prevent nearby pieces of fluid from acquiring significantly different velocities, and the flow becomes turbulent. When Re $\ll$ 1, the viscous term is large, and viscosity is the dominant physical effect. The flow oozes, as when pouring cold honey.

3.5 Predicting the period of a pendulum

Lumping not only turns integration into multiplication, it turns nonlinear into linear differential equations. Our example is the analysis of the period of a pendulum, for centuries the basis of Western timekeeping.

▶ *How does the period of a pendulum depend on its amplitude?*

The amplitude θ_0 is the maximum angle of the swing; for a lossless pendulum released from rest, it is also the angle of release. The effect of amplitude is contained in the solution to the pendulum differential equation (see [24] for the equation's derivation):

$$\frac{d^2\theta}{dt^2} + \frac{g}{l} \sin \theta = 0. \tag{3.32}$$

The analysis will use all our tools: dimensions (Section 3.5.2), easy cases (Section 3.5.1 and Section 3.5.3), and lumping (Section 3.5.4).

3.5 Predicting the period of a pendulum

Problem 3.23 Angles
Explain why angles are dimensionless.

Problem 3.24 Checking and using dimensions
Does the pendulum equation have correct dimensions? Use dimensional analysis to show that the equation cannot contain the mass of the bob (except as a common factor that divides out).

3.5.1 Small amplitudes: Applying extreme cases

The pendulum equation is difficult because of its nonlinear factor $\sin\theta$. Fortunately, the factor is easy in the small-amplitude extreme case $\theta \to 0$. In that limit, the height of the triangle, which is $\sin\theta$, is almost exactly the arclength θ. Therefore, for small angles, $\sin\theta \approx \theta$.

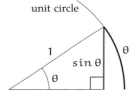

Problem 3.25 Chord approximation
The $\sin\theta \approx \theta$ approximation replaces the arc with a straight, vertical line. To make a more accurate approximation, replace the arc with the chord (a straight but nonvertical line). What is the resulting approximation for $\sin\theta$?

In the small-amplitude extreme, the pendulum equation becomes linear:

$$\frac{d^2\theta}{dt^2} + \frac{g}{l}\theta = 0. \tag{3.33}$$

Compare this equation to the spring–mass equation (Section 3.4)

$$\frac{d^2x}{dt^2} + \frac{k}{m}x = 0. \tag{3.34}$$

The equations correspond with x analogous to θ and k/m analogous to g/l. The frequency of the spring–mass system is $\omega = \sqrt{k/m}$, and its period is $T = 2\pi/\omega = 2\pi\sqrt{m/k}$. For the pendulum equation, the corresponding period is

$$T = 2\pi\sqrt{\frac{l}{g}} \quad \text{(for small amplitudes).} \tag{3.35}$$

(This analysis is a preview of the method of analogy, which is the subject of Chapter 6.)

Problem 3.26 Checking dimensions
Does the period $2\pi\sqrt{l/g}$ have correct dimensions?

Problem 3.27 Checking extreme cases
Does the period $T = 2\pi\sqrt{l/g}$ make sense in the extreme cases $g \to \infty$ and $g \to 0$?

Problem 3.28 Possible coincidence
Is it a coincidence that $g \approx \pi^2 \, \mathrm{m\,s^{-2}}$? (For an extensive historical discussion that involves the pendulum, see [1] and more broadly also [4, 27, 42].)

Problem 3.29 Conical pendulum for the constant
The dimensionless factor of 2π can be derived using an insight from Huygens [15, p. 79]: to analyze the motion of a pendulum moving in a horizontal circle (a conical pendulum). Projecting its two-dimensional motion onto a vertical screen produces one-dimensional pendulum motion, so the period of the two-dimensional motion is the same as the period of one-dimensional pendulum motion! Use that idea along with Newton's laws of motion to explain the 2π.

3.5.2 Arbitrary amplitudes: Applying dimensional analysis

The preceding results might change if the amplitude θ_0 is no longer small.

▶ *As θ_0 increases, does the period increase, remain constant, or decrease?*

Any analysis becomes cleaner if expressed using dimensionless groups (Section 2.4.1). This problem involves the period T, length l, gravitational strength g, and amplitude θ_0. Therefore, T can belong to the dimensionless group $T/\sqrt{l/g}$. Because angles are dimensionless, θ_0 is itself a dimensionless group. The two groups $T/\sqrt{l/g}$ and θ_0 are independent and fully describe the problem (Problem 3.30).

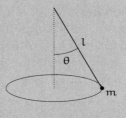

An instructive contrast is the ideal spring–mass system. The period T, spring constant k, and mass m can form the dimensionless group $T/\sqrt{m/k}$; but the amplitude x_0, as the only quantity containing a length, cannot be part of any dimensionless group (Problem 3.20) and cannot therefore affect the period of the spring–mass system. In contrast,

3.5 Predicting the period of a pendulum

the pendulum's amplitude θ_0 is already a dimensionless group, so it can affect the period of the system.

> **Problem 3.30 Choosing dimensionless groups**
> Check that period T, length l, gravitational strength g, and amplitude θ_0 produce two independent dimensionless groups. In constructing useful groups for analyzing the period, why should T appear in only one group? And why should θ_0 not appear in the same group as T?

Two dimensionless groups produce the general dimensionless form

$$\text{one group} = \text{function of the other group}, \tag{3.36}$$

so

$$\frac{T}{\sqrt{l/g}} = \text{function of } \theta_0. \tag{3.37}$$

Because $T/\sqrt{l/g} = 2\pi$ when $\theta_0 = 0$ (the small-amplitude limit), factor out the 2π to simplify the subsequent equations, and define a dimensionless period h as follows:

$$\frac{T}{\sqrt{l/g}} = 2\pi\, h(\theta_0). \tag{3.38}$$

The function h contains all information about how amplitude affects the period of a pendulum. Using h, the original question about the period becomes the following: Is h an increasing, constant, or decreasing function of amplitude? This question is answered in the following section.

3.5.3 Large amplitudes: Extreme cases again

For guessing the general behavior of h as a function of amplitude, useful clues come from evaluating h at two amplitudes. One easy amplitude is the extreme of zero amplitude, where $h(0) = 1$. A second easy amplitude is the opposite extreme of large amplitudes.

▶ *How does the period behave at large amplitudes? As part of that question, what is a large amplitude?*

An interesting large amplitude is $\pi/2$, which means releasing the pendulum from horizontal. However, at $\pi/2$ the exact h is the following awful expression (Problem 3.31):

$$h(\pi/2) = \frac{\sqrt{2}}{\pi} \int_0^{\pi/2} \frac{d\theta}{\sqrt{\cos\theta}}. \tag{3.39}$$

Is this integral less than, equal to, or more than 1? Who knows? The integral is likely to have no closed form and to require numerical evaluation (Problem 3.32).

> **Problem 3.31   General expression for h**
> Use conservation of energy to show that the period is
> $$T(\theta_0) = 2\sqrt{2}\sqrt{\frac{l}{g}} \int_0^{\theta_0} \frac{d\theta}{\sqrt{\cos\theta - \cos\theta_0}}. \tag{3.40}$$
> Confirm that the equivalent dimensionless statement is
> $$h(\theta_0) = \frac{\sqrt{2}}{\pi} \int_0^{\theta_0} \frac{d\theta}{\sqrt{\cos\theta - \cos\theta_0}}. \tag{3.41}$$
> For horizontal release, $\theta_0 = \pi/2$, and
> $$h(\pi/2) = \frac{\sqrt{2}}{\pi} \int_0^{\pi/2} \frac{d\theta}{\sqrt{\cos\theta}}. \tag{3.42}$$

> **Problem 3.32   Numerical evaluation for horizontal release**
> Why do the lumping recipes (Section 3.2) fail for the integrals in Problem 3.31? Compute $h(\pi/2)$ using numerical integration.

Because $\theta_0 = \pi/2$ is not a helpful extreme, be even more extreme. Try $\theta_0 = \pi$, which means releasing the pendulum bob from vertical. If the bob is connected to the pivot point by a string, however, a vertical release would mean that the bob falls straight down instead of oscillating. This novel behavior is neither included in nor described by the pendulum differential equation.

Fortunately, a thought experiment is cheap to improve: Replace the string with a massless steel rod. Balanced perfectly at $\theta_0 = \pi$, the pendulum bob hangs upside down forever, so $T(\pi) = \infty$ and $h(\pi) = \infty$. Thus, $h(\pi) > 1$ and $h(0) = 1$. From these data, the most likely conjecture is that h increases monotonically with amplitude. Although h could first decrease and then increase, such twists and turns would be surprising behavior from such a clean differential equation. (For the behavior of h near $\theta_0 = \pi$, see Problem 3.34.)

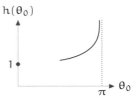

3.5 Predicting the period of a pendulum

Problem 3.33 Small but nonzero amplitude
As the amplitude approaches π, the dimensionless period h diverges to infinity; at zero amplitude, h = 1. But what about the derivative of h? At zero amplitude ($\theta_0 = 0$), does $h(\theta_0)$ have zero slope (curve A) or positive slope (curve B)?

Problem 3.34 Nearly vertical release
Imagine releasing the pendulum from almost vertical: an initial angle $\pi - \beta$ with β tiny. As a function of β, roughly how long does the pendulum take to rotate by a significant angle—say, by 1 rad? Use that information to predict how $h(\theta_0)$ behaves when $\theta_0 \approx \pi$. Check and refine your conjectures using the tabulated values. Then predict $h(\pi - 10^{-5})$.

β	$h(\pi - \beta)$
10^{-1}	2.791297
10^{-2}	4.255581
10^{-3}	5.721428
10^{-4}	7.187298

3.5.4 Moderate amplitudes: Applying lumping

The conjecture that h increases monotonically was derived using the extremes of zero and vertical amplitude, so it should apply at intermediate amplitudes. Before taking that statement on faith, recall a proverb from arms-control negotiations: "Trust, but verify."

▷ *At moderate (small but nonzero) amplitudes, does the period, or its dimensionless cousin h, increase with amplitude?*

In the zero-amplitude extreme, $\sin \theta$ is close to θ. That approximation turned the nonlinear pendulum equation

$$\frac{d^2\theta}{dt^2} + \frac{g}{l}\sin\theta = 0 \qquad (3.43)$$

into the linear, ideal-spring equation—in which the period is independent of amplitude.

At nonzero amplitude, however, θ and $\sin \theta$ differ and their difference affects the period. To account for the difference and predict the period, split $\sin \theta$ into the tractable factor θ and an adjustment factor $f(\theta)$. The resulting equation is

$$\frac{d^2\theta}{dt^2} + \frac{g}{l}\theta \underbrace{\frac{\sin\theta}{\theta}}_{f(\theta)} = 0. \qquad (3.44)$$

The nonconstant f(θ) encapsulates the nonlinearity of
the pendulum equation. When θ is tiny, f(θ) ≈ 1: The
pendulum behaves like a linear, ideal-spring system.
But when θ is large, f(θ) falls significantly below 1,
making the ideal-spring approximation significantly
inaccurate. As is often the case, a changing process is
difficult to analyze—for example, see the awful integrals in Problem 3.31.
As a countermeasure, make a lumping approximation by replacing the
changing f(θ) with a constant.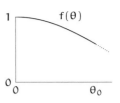

The simplest constant is f(0). Then the pendulum differential equation becomes

$$\frac{d^2\theta}{dt^2} + \frac{g}{l}\theta = 0. \qquad (3.45)$$

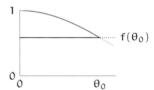

This equation is, again, the ideal-spring equation.
In this approximation, period does not depend on amplitude, so $h = 1$ for
all amplitudes. For determining how the period of an unapproximated
pendulum depends on amplitude, the $f(\theta) \to f(0)$ lumping approximation discards too much information.

Therefore, replace f(θ) with the other extreme
$f(\theta_0)$. Then the pendulum equation becomes

$$\frac{d^2\theta}{dt^2} + \frac{g}{l}\theta f(\theta_0) = 0. \qquad (3.46)$$

▶ *Is this equation linear? What physical system does it describe?*

Because $f(\theta_0)$ is a constant, this equation is linear! It describes a zero-amplitude pendulum on a planet with gravity g_{eff} that is slightly weaker than earth gravity—as shown by the following slight regrouping:

$$\frac{d^2\theta}{dt^2} + \overbrace{\frac{gf(\theta_0)}{l}}^{g_{\text{eff}}}\theta = 0. \qquad (3.47)$$

Because the zero-amplitude pendulum has period $T = 2\pi\sqrt{l/g}$, the zero-amplitude, low-gravity pendulum has period

$$T(\theta_0) \approx 2\pi\sqrt{\frac{l}{g_{\text{eff}}}} = 2\pi\sqrt{\frac{l}{gf(\theta_0)}}. \qquad (3.48)$$

3.5 Predicting the period of a pendulum

Using the dimensionless period h avoids writing the factors of 2π, l, and g, and it yields the simple prediction

$$h(\theta_0) \approx f(\theta_0)^{-1/2} = \left(\frac{\sin \theta_0}{\theta_0}\right)^{-1/2}. \qquad (3.49)$$

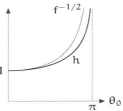

At moderate amplitudes the approximation closely follows the exact dimensionless period (dark curve). As a bonus, it also predicts $h(\pi) = \infty$, so it agrees with the thought experiment of releasing the pendulum from upright (Section 3.5.3).

▶ *How much larger than the period at zero amplitude is the period at 10° amplitude?*

A 10° amplitude is roughly 0.17 rad, a moderate angle, so the approximate prediction for h can itself accurately be approximated using a Taylor series. The Taylor series for $\sin \theta$ begins $\theta - \theta^3/6$, so

$$f(\theta_0) = \frac{\sin \theta_0}{\theta_0} \approx 1 - \frac{\theta_0^2}{6}. \qquad (3.50)$$

Then $h(\theta_0)$, which is roughly $f(\theta_0)^{-1/2}$, becomes

$$h(\theta_0) \approx \left(1 - \frac{\theta_0^2}{6}\right)^{-1/2}. \qquad (3.51)$$

Another Taylor series yields $(1+x)^{-1/2} \approx 1 - x/2$ (for small x). Therefore,

$$h(\theta_0) \approx 1 + \frac{\theta_0^2}{12}. \qquad (3.52)$$

Restoring the dimensioned quantities gives the period itself.

$$T \approx 2\pi \sqrt{\frac{l}{g}} \left(1 + \frac{\theta_0^2}{12}\right). \qquad (3.53)$$

Compared to the period at zero amplitude, a 10° amplitude produces a fractional increase of roughly $\theta_0^2/12 \approx 0.0025$ or 0.25%. Even at moderate amplitudes, the period is nearly independent of amplitude!

> **Problem 3.35 Slope revisited**
> Use the preceding result for $h(\theta_0)$ to check your conclusion in Problem 3.33 about the slope of $h(\theta_0)$ at $\theta_0 = 0$.

▶ *Does our lumping approximation underestimate or overestimate the period?*

The lumping approximation simplified the pendulum differential equation by replacing $f(\theta)$ with $f(\theta_0)$. Equivalently, it assumed that the mass always remained at the endpoints of the motion where $|\theta| = \theta_0$. Instead, the pendulum spends much of its time at intermediate positions where $|\theta| < \theta_0$ and $f(\theta) > f(\theta_0)$. Therefore, the average f is greater than $f(\theta_0)$. Because h is inversely related to f ($h = f^{-1/2}$), the $f(\theta) \to f(\theta_0)$ lumping approximation overestimates h and the period.

The $f(\theta) \to f(0)$ lumping approximation, which predicts $T = 2\pi\sqrt{l/g}$, underestimates the period. Therefore, the true coefficient of the θ_0^2 term in the period approximation

$$T \approx 2\pi\sqrt{\frac{l}{g}}\left(1 + \frac{\theta_0^2}{12}\right) \tag{3.54}$$

lies between 0 and 1/12. A natural guess is that the coefficient lies halfway between these extremes—namely, 1/24. However, the pendulum spends more time toward the extremes (where $f(\theta) = f(\theta_0)$) than it spends near the equilibrium position (where $f(\theta) = f(0)$). Therefore, the true coefficient is probably closer to 1/12—the prediction of the $f(\theta) \to f(\theta_0)$ approximation—than it is to 0. An improved guess might be two-thirds of the way from 0 to 1/12, namely 1/18.

In comparison, a full successive-approximation solution of the pendulum differential equation gives the following period [13, 33]:

$$T = 2\pi\sqrt{\frac{l}{g}}\left(1 + \frac{1}{16}\theta_0^2 + \frac{11}{3072}\theta_0^4 + \cdots\right). \tag{3.55}$$

Our educated guess of 1/18 is very close to the true coefficient of 1/16!

3.6 Summary and further problems

Lumping turns calculus on its head. Whereas calculus analyzes a changing process by dividing it into ever finer intervals, lumping simplifies a changing process by combining it into one unchanging process. It turns curves into straight lines, difficult integrals into multiplication, and mildly nonlinear differential equations into linear differential equations.

> ...*the crooked shall be made straight, and the rough places plain.* (Isaiah 40:4)

3.6 Summary and further problems

Problem 3.36 FWHM for another decaying function
Use the FWHM heuristic to estimate

$$\int_{-\infty}^{\infty} \frac{dx}{1+x^4}. \tag{3.56}$$

Then compare the estimate with the exact value of $\pi/\sqrt{2}$. For an enjoyable additional problem, derive the exact value.

Problem 3.37 Hypothetical pendulum equation
Suppose the pendulum equation had been

$$\frac{d^2\theta}{d\theta^2} + \frac{g}{l}\tan\theta = 0. \tag{3.57}$$

How would the period T depend on amplitude θ_0? In particular, as θ_0 increases, would T decrease, remain constant, or increase? What is the slope $dT/d\theta_0$ at zero amplitude? Compare your results with the results of Problem 3.33.

For small but nonzero θ_0, find an approximate expression for the dimensionless period $h(\theta_0)$ and use it to check your previous conclusions.

Problem 3.38 Gaussian 1-sigma tail
The Gaussian probability density function with zero mean and unit variance is

$$p(x) = \frac{e^{-x^2/2}}{\sqrt{2\pi}}. \tag{3.58}$$

The area of its tail is an important quantity in statistics, but it has no closed form. In this problem you estimate the area of the 1-sigma tail

$$\int_1^\infty \frac{e^{-x^2/2}}{\sqrt{2\pi}}\, dx. \tag{3.59}$$

a. Sketch the above Gaussian and shade the 1-sigma tail.
b. Use the $1/e$ lumping heuristic (Section 3.2.1) to estimate the area.
c. Use the FWHM heuristic to estimate the area.
d. Compare the two lumping estimates with the result of numerical integration:

$$\int_1^\infty \frac{e^{-x^2/2}}{\sqrt{2\pi}}\, dx = \frac{1-\mathrm{erf}(1/\sqrt{2})}{2} \approx 0.159, \tag{3.60}$$

where $\mathrm{erf}(z)$ is the error function.

Problem 3.39 Distant Gaussian tails
For the canonical probability Gaussian, estimate the area of its n-sigma tail (for large n). In other words, estimate

$$\int_n^\infty \frac{e^{-x^2/2}}{\sqrt{2\pi}}\, dx. \tag{3.61}$$

4
Pictorial proofs

4.1	Adding odd numbers	58
4.2	Arithmetic and geometric means	60
4.3	Approximating the logarithm	66
4.4	Bisecting a triangle	70
4.5	Summing series	73
4.6	*Summary and further problems*	75

Have you ever worked through a proof, understood and confirmed each step, yet still not believed the theorem? You realize *that* the theorem is true, but not *why* it is true.

To see the same contrast in a familiar example, imagine learning that your child has a fever and hearing the temperature in Fahrenheit or Celsius degrees, whichever is less familiar. In my everyday experience, temperatures are mostly in Fahrenheit. When I hear about a temperature of 40°C, I therefore react in two stages:

1. I convert 40°C to Fahrenheit: $40 \times 1.8 + 32 = 104$.
2. I react: "Wow, 104°F. That's dangerous! Get thee to a doctor!"

The Celsius temperature, although symbolically equivalent to the Fahrenheit temperature, elicits no reaction. My danger sense activates only after the temperature conversion connects the temperature to my experience.

A symbolic description, whether a proof or an unfamiliar temperature, is unconvincing compared to an argument that speaks to our perceptual system. The reason lies in how our brains acquired the capacity for symbolic reasoning. (See *Evolving Brains* [2] for an illustrated, scholarly history of the brain.) Symbolic, sequential reasoning requires language, which has evolved for only 10^5 yr. Although 10^5 yr spans many human lifetimes, it is an evolutionary eyeblink. In particular, it is short compared to the time span over which our perceptual hardware has evolved: For several hundred million years, organisms have refined their capacities for hearing, smelling, tasting, touching, and seeing.

Evolution has worked 1000 times longer on our perceptual abilities than on our symbolic-reasoning abilities. Compared to our perceptual hardware, our symbolic, sequential hardware is an ill-developed latecomer. Not surprisingly, our perceptual abilities far surpass our symbolic abilities. Even an apparently high-level symbolic activity such as playing grandmaster chess uses mostly perceptual hardware [16]. *Seeing* an idea conveys to us a depth of understanding that a symbolic description of it cannot easily match.

> **Problem 4.1 Computers versus people**
> At tasks like expanding $(x+2y)^{50}$, computers are much faster than people. At tasks like recognizing faces or smells, even young children are much faster than current computers. How do you explain these contrasts?
>
> **Problem 4.2 Linguistic evidence for the importance of perception**
> In your favorite language(s), think of the many sensory synonyms for understanding (for example, grasping).

4.1 Adding odd numbers

To illustrate the value of pictures, let's find the sum of the first n odd numbers (also the subject of Problem 2.25):

$$S_n = \underbrace{1 + 3 + 5 + \cdots + (2n-1)}_{n \text{ terms}}. \tag{4.1}$$

Easy cases such as $n = 1$, 2, or 3 lead to the conjecture that $S_n = n^2$. But how can the conjecture be proved? The standard symbolic method is proof by induction:

1. Verify that $S_n = n^2$ for the *base case* $n = 1$. In that case, S_1 is 1, as is n^2, so the base case is verified.

2. Make the *induction hypothesis:* Assume that $S_m = m^2$ for m less than or equal to a maximum value n. For this proof, the following, weaker induction hypothesis is sufficient:

4.1 Adding odd numbers

$$\sum_{1}^{n}(2k-1) = n^2. \tag{4.2}$$

In other words, we assume the theorem only in the case that $m = n$.

3. Perform the *induction step*: Use the induction hypothesis to show that $S_{n+1} = (n+1)^2$. The sum S_{n+1} splits into two pieces:

$$S_{n+1} = \sum_{1}^{n+1}(2k-1) = (2n+1) + \sum_{1}^{n}(2k-1). \tag{4.3}$$

Thanks to the induction hypothesis, the sum on the right is n^2. Thus

$$S_{n+1} = (2n+1) + n^2, \tag{4.4}$$

which is $(n+1)^2$; and the theorem is proved.

Although these steps prove the theorem, *why* the sum S_n ends up as n^2 still feels elusive.

That missing understanding—the kind of gestalt insight described by Wertheimer [48]—requires a pictorial proof. Start by drawing each odd number as an L-shaped puzzle piece:

$$\tag{4.5}$$

▶ *How do these pieces fit together?*

Then compute S_n by fitting together the puzzle pieces as follows:

$$\tag{4.6}$$

Each successive odd number—each piece—extends the square by 1 unit in height and width, so the n terms build an $n \times n$ square. [Or is it an $(n-1) \times (n-1)$ square?] Therefore, their sum is n^2. After grasping this pictorial proof, you cannot forget why adding up the first n odd numbers produces n^2.

> **Problem 4.3   Triangular numbers**
> Draw a picture or pictures to show that
> $$1+2+3+\cdots+n+\cdots+3+2+1 = n^2. \qquad (4.7)$$
> Then show that
> $$1+2+3+\cdots+n = \frac{n(n+1)}{2}. \qquad (4.8)$$
>
> **Problem 4.4   Three dimensions**
> Draw a picture to show that
> $$\sum_{0}^{n}(3k^2+3k+1) = (n+1)^3. \qquad (4.9)$$
>
> Give pictorial explanations for the 1 in the summand $3k^2+3k+1$; for the 3 and the k^2 in $3k^2$; and for the 3 and the k in 3k.

4.2 Arithmetic and geometric means

The next pictorial proof starts with two nonnegative numbers—for example, 3 and 4—and compares the following two averages:

$$\text{arithmetic mean} \equiv \frac{3+4}{2} = 3.5; \qquad (4.10)$$

$$\text{geometric mean} \equiv \sqrt{3\times 4} \approx 3.464. \qquad (4.11)$$

Try another pair of numbers—for example, 1 and 2. The arithmetic mean is 1.5; the geometric mean is $\sqrt{2} \approx 1.414$. For both pairs, the geometric mean is smaller than the arithmetic mean. This pattern is general; it is the famous arithmetic-mean–geometric-mean (AM–GM) inequality [18]:

$$\underbrace{\frac{a+b}{2}}_{\text{AM}} \geqslant \underbrace{\sqrt{ab}}_{\text{GM}}. \qquad (4.12)$$

(The inequality requires that $a, b \geqslant 0$.)

> **Problem 4.5   More numerical examples**
> Test the AM–GM inequality using varied numerical examples. What do you notice when a and b are close to each other? Can you formalize the pattern? (See also Problem 4.16.)

4.2 Arithmetic and geometric means

4.2.1 Symbolic proof

The AM–GM inequality has a pictorial and a symbolic proof. The symbolic proof begins with $(a-b)^2$—a surprising choice because the inequality contains $a+b$ rather than $a-b$. The second odd choice is to form $(a-b)^2$. It is nonnegative, so $a^2 - 2ab + b^2 \geq 0$. Now magically decide to add $4ab$ to both sides. The result is

$$\underbrace{a^2 + 2ab + b^2}_{(a+b)^2} \geq 4ab. \tag{4.13}$$

The left side is $(a+b)^2$, so $a+b \geq 2\sqrt{ab}$ and

$$\frac{a+b}{2} \geq \sqrt{ab}. \tag{4.14}$$

Although each step is simple, the whole chain seems like magic and leaves the *why* mysterious. If the algebra had ended with $(a+b)/4 \geq \sqrt{ab}$, it would not look obviously wrong. In contrast, a convincing proof would leave us feeling that the inequality cannot help but be true.

4.2.2 Pictorial proof

This satisfaction is provided by a pictorial proof.

▶ *What is pictorial, or geometric, about the geometric mean?*

A geometric picture for the geometric mean starts with a right triangle. Lay it with its hypotenuse horizontal; then cut it with the altitude x into the light and dark subtriangles. The hypotenuse splits into two lengths a and b, and the altitude x is their geometric mean $\sqrt{ab}$.

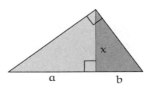

▶ *Why is the altitude x equal to $\sqrt{ab}$?*

To show that $x = \sqrt{ab}$, compare the small, dark triangle to the large, light triangle by rotating the small triangle and laying it on the large triangle. The two triangles are similar! Therefore, their aspect ratios (the ratio of the short to the long side) are identical. In symbols, $x/a = b/x$: The altitude x is therefore the geometric mean $\sqrt{ab}$.

The uncut right triangle represents the geometric-mean portion of the AM–GM inequality. The arithmetic mean $(a+b)/2$ also has a picture, as one-half of the hypotenuse. Thus, the inequality claims that

$$\frac{\text{hypotenuse}}{2} \geqslant \text{altitude}. \tag{4.15}$$

Alas, this claim is not pictorially obvious.

▶ *Can you find an alternative geometric interpretation of the arithmetic mean that makes the AM–GM inequality pictorially obvious?*

The arithmetic mean is also the radius of a circle with diameter $a + b$. Therefore, circumscribe a semicircle around the triangle, matching the circle's diameter with the hypotenuse $a + b$ (Problem 4.7). The altitude cannot exceed the radius; therefore,

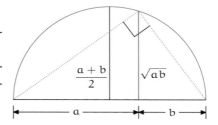

$$\frac{a+b}{2} \geqslant \sqrt{ab}. \tag{4.16}$$

Furthermore, the two sides are equal only when the altitude of the triangle is also a radius of the semicircle—namely when $a = b$. The picture therefore contains the inequality and its equality condition in one easy-to-grasp object. (An alternative pictorial proof of the AM–GM inequality is developed in Problem 4.33.)

Problem 4.6 Circumscribing a circle around a triangle
Here are a few examples showing a circle circumscribed around a triangle.

Draw a picture to show that the circle is uniquely determined by the triangle.

Problem 4.7 Finding the right semicircle
A triangle uniquely determines its circumscribing circle (Problem 4.6). However, the circle's diameter might not align with a side of the triangle. Can a semicircle always be circumscribed around a right triangle while aligning the circle's diameter along the hypotenuse?

4.2 Arithmetic and geometric means

> **Problem 4.8   Geometric mean of three numbers**
> For three nonnegative numbers, the AM–GM inequality is
> $$\frac{a+b+c}{3} \geqslant (abc)^{1/3}. \tag{4.17}$$
> Why is this inequality, in contrast to its two-number cousin, unlikely to have a geometric proof? (If you find a proof, let me know.)

4.2.3 Applications

Arithmetic and geometric means have wide mathematical application. The first application is a problem more often solved with derivatives: Fold a fixed length of fence into a rectangle enclosing the largest garden.

▶ *What shape of rectangle maximizes the area?*

The problem involves two quantities: a perimeter that is fixed and an area to maximize. If the perimeter is related to the arithmetic mean and the area to the geometric mean, then the AM–GM inequality might help maximize the area. The perimeter $P = 2(a+b)$ is four times the arithmetic mean, and the area $A = ab$ is the square of the geometric mean. Therefore, from the AM–GM inequality,

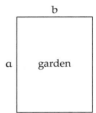

$$\underbrace{\frac{P}{4}}_{\text{AM}} \geqslant \underbrace{\sqrt{A}}_{\text{GM}} \tag{4.18}$$

with equality when $a = b$. The left side is fixed by the amount of fence. Thus the right side, which varies depending on a and b, has a maximum of $P/4$ when $a = b$. The maximal-area rectangle is a square.

> **Problem 4.9   Direct pictorial proof**
> The AM–GM reasoning for the maximal rectangular garden is indirect pictorial reasoning. It is symbolic reasoning built upon the pictorial proof for the AM–GM inequality. Can you draw a picture to show directly that the square is the optimal shape?
>
> **Problem 4.10   Three-part product**
> Find the maximum value of $f(x) = x^2(1-2x)$ for $x \geqslant 0$, without using calculus. Sketch $f(x)$ to confirm your answer.

Problem 4.11   Unrestricted maximal area
If the garden need not be rectangular, what is the maximal-area shape?

Problem 4.12   Volume maximization
Build an open-topped box as follows: Start with a unit square, cut out four identical corners, and fold in the flaps. The box has volume $V = x(1 - 2x)^2$, where x is the side length of a corner cutout. What choice of x maximizes the volume of the box?

Here is a plausible analysis modeled on the analysis of the rectangular garden. Set $a = x$, $b = 1 - 2x$, and $c = 1 - 2x$. Then abc is the volume V, and $V^{1/3} = \sqrt[3]{abc}$ is the geometric mean (Problem 4.8). Because the geometric mean never exceeds the arithmetic mean and because the two means are equal when $a = b = c$, the maximum volume is attained when $x = 1 - 2x$. Therefore, choosing $x = 1/3$ should maximize the volume of the box.

Now show that this choice is wrong by graphing $V(x)$ or setting $dV/dx = 0$; explain what is wrong with the preceding reasoning; and make a correct version.

Problem 4.13   Trigonometric minimum
Find the minimum value of

$$\frac{9x^2 \sin^2 x + 4}{x \sin x} \qquad (4.19)$$

in the region $x \in (0, \pi)$.

Problem 4.14   Trigonometric maximum
In the region $t \in [0, \pi/2]$, maximize $\sin 2t$ or, equivalently, $2 \sin t \cos t$.

The second application of arithmetic and geometric means is a modern, amazingly rapid method for computing π [5, 6]. Ancient methods for computing π included calculating the perimeter of many-sided regular polygons and provided a few decimal places of accuracy.

Recent computations have used Leibniz's arctangent series

$$\arctan x = x - \frac{x^3}{3} + \frac{x^5}{5} - \frac{x^7}{7} + \cdots. \qquad (4.20)$$

Imagine that you want to compute π to 10^9 digits, perhaps to test the hardware of a new supercomputer or to study whether the digits of π are random (a theme in Carl Sagan's novel *Contact* [40]). Setting $x = 1$ in the Leibniz series produces $\pi/4$, but the series converges extremely slowly. Obtaining 10^9 digits requires roughly 10^{10^9} terms—far more terms than atoms in the universe.

4.2 Arithmetic and geometric means

Fortunately, a surprising trigonometric identity due to John Machin (1686–1751)

$$\arctan 1 = 4\arctan\frac{1}{5} - \arctan\frac{1}{239} \qquad (4.21)$$

accelerates the convergence by reducing x:

$$\frac{\pi}{4} = 4 \times \underbrace{\left(1 - \frac{1}{3 \times 5^3} + \cdots\right)}_{\arctan(1/5)} - \underbrace{\left(1 - \frac{1}{3 \times 239^3} + \cdots\right)}_{\arctan(1/239)}. \qquad (4.22)$$

Even with the speedup, 10^9-digit accuracy requires calculating roughly 10^9 terms.

In contrast, the modern Brent–Salamin algorithm [3, 41], which relies on arithmetic and geometric means, converges to π extremely rapidly. The algorithm is closely related to amazingly accurate methods for calculating the perimeter of an ellipse (Problem 4.15) and also for calculating mutual inductance [23]. The algorithm generates several sequences by starting with $a_0 = 1$ and $g_0 = 1/\sqrt{2}$; it then computes successive arithmetic means a_n, geometric means g_n, and their squared differences d_n.

$$a_{n+1} = \frac{a_n + g_n}{2}, \qquad g_{n+1} = \sqrt{a_n g_n}, \qquad d_n = a_n^2 - g_n^2. \qquad (4.23)$$

The a and g sequences rapidly converge to a number $M(a_0, g_0)$ called the arithmetic–geometric mean of a_0 and g_0. Then $M(a_0, g_0)$ and the difference sequence d determine π.

$$\pi = \frac{4M(a_0, g_0)^2}{1 - \sum_{j=1}^{\infty} 2^{j+1} d_j}. \qquad (4.24)$$

The d sequence approaches zero quadratically; in other words, $d_{n+1} \sim d_n^2$ (Problem 4.16). Therefore, each iteration in this computation of π doubles the digits of accuracy. A billion-digit calculation of π requires only about 30 iterations—far fewer than the 10^{10^9} terms using the arctangent series with $x = 1$ or even than the 10^9 terms using Machin's speedup.

Problem 4.15 Perimeter of an ellipse

To compute the perimeter of an ellipse with semimajor axis a_0 and semiminor axis g_0, compute the a, g, and d sequences and the common limit $M(a_0, g_0)$ of the a and g sequences, as for the computation of π. Then the perimeter P can be computed with the following formula:

$$P = \frac{A}{M(a_0, g_0)} \left(a_0^2 - B \sum_{j=0}^{\infty} 2^j d_j \right), \qquad (4.25)$$

where A and B are constants for you to determine. Use the method of easy cases (Chapter 2) to determine their values. (See [3] to check your values and for a proof of the completed formula.)

Problem 4.16   Quadratic convergence
Start with $a_0 = 1$ and $g_0 = 1/\sqrt{2}$ (or any other positive pair) and follow several iterations of the AM–GM sequence

$$a_{n+1} = \frac{a_n + g_n}{2} \quad \text{and} \quad g_{n+1} = \sqrt{a_n g_n}. \qquad (4.26)$$

Then generate $d_n = a_n^2 - g_n^2$ and $\log_{10} d_n$ to check that $d_{n+1} \sim d_n^2$ (quadratic convergence).

Problem 4.17   Rapidity of convergence
Pick a positive x_0; then generate a sequence by the iteration

$$x_{n+1} = \frac{1}{2}\left(x_n + \frac{2}{x_n}\right) \quad (n \geq 0). \qquad (4.27)$$

To what and how rapidly does the sequence converge? What if $x_0 < 0$?

4.3 Approximating the logarithm

A function is often approximated by its Taylor series

$$f(x) = f(0) + x \left.\frac{df}{dx}\right|_{x=0} + \frac{x^2}{2} \left.\frac{d^2f}{dx^2}\right|_{x=0} + \cdots, \qquad (4.28)$$

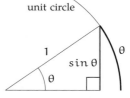

which looks like an unintuitive sequence of symbols. Fortunately, pictures often explain the first and most important terms in a function approximation. For example, the one-term approximation $\sin\theta \approx \theta$, which replaces the altitude of the triangle by the arc of the circle, turns the nonlinear pendulum differential equation into a tractable, linear equation (Section 3.5).

Another Taylor-series illustration of the value of pictures come from the series for the logarithm function:

$$\ln(1+x) = x - \frac{x^2}{2} + \frac{x^3}{3} - \cdots. \qquad (4.29)$$

4.3 Approximating the logarithm

Its first term, x, will lead to the wonderful approximation $(1+x)^n \approx e^{nx}$ for small x and arbitrary n (Section 5.3.4). Its second term, $-x^2/2$, helps evaluate the accuracy of that approximation. These first two terms are the most useful terms—and they have pictorial explanations.

The starting picture is the integral representation

$$\ln(1+x) = \int_0^x \frac{dt}{1+t}. \qquad (4.30)$$

▶ *What is the simplest approximation for the shaded area?*

As a first approximation, the shaded area is roughly the circumscribed rectangle—an example of lumping. The rectangle has area x:

$$\text{area} = \underbrace{\text{height}}_{1} \times \underbrace{\text{width}}_{x} = x. \qquad (4.31)$$

This area reproduces the first term in the Taylor series. Because it uses a circumscribed rectangle, it slightly overestimates $\ln(1+x)$.

The area can also be approximated by drawing an inscribed rectangle. Its width is again x, but its height is not 1 but rather $1/(1+x)$, which is approximately $1-x$ (Problem 4.18). Thus the inscribed rectangle has the approximate area $x(1-x) = x - x^2$. This area slightly underestimates $\ln(1+x)$.

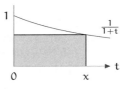

> **Problem 4.18 Picture for approximating the reciprocal function**
> Confirm the approximation
> $$\frac{1}{1+x} \approx 1-x \quad \text{(for small x)} \qquad (4.32)$$
> by trying $x = 0.1$ or $x = 0.2$. Then draw a picture to illustrate the equivalent approximation $(1-x)(1+x) \approx 1$.

We now have two approximations to $\ln(1+x)$. The first and slightly simpler approximation came from drawing the circumscribed rectangle. The second approximation came from drawing the inscribed rectangle. Both dance around the exact value.

▶ *How can the inscribed- and circumscribed-rectangle approximations be combined to make an improved approximation?*

One approximation overestimates the area, and the other underestimates the area; their average ought to improve on either approximation. The average is a trapezoid with area

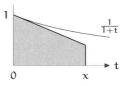

$$\frac{x + (x - x^2)}{2} = x - \frac{x^2}{2}. \qquad (4.33)$$

This area reproduces the first two terms of the full Taylor series

$$\ln(1 + x) = x - \frac{x^2}{2} + \frac{x^3}{3} - \cdots. \qquad (4.34)$$

> **Problem 4.19 Cubic term**
> Estimate the cubic term in the Taylor series by estimating the difference between the trapezoid and the true area.

For these logarithm approximations, the hardest problem is $\ln 2$.

$$\ln(1 + 1) \approx \begin{cases} 1 & \text{(one term)} \\ 1 - \frac{1}{2} & \text{(two terms)}. \end{cases} \qquad (4.35)$$

Both approximations differ significantly from the true value (roughly 0.693). Even moderate accuracy for $\ln 2$ requires many terms of the Taylor series, far beyond what pictures explain (Problem 4.20). The problem is that x in $\ln(1 + x)$ is 1, so the x^n factor in each term of the Taylor series does not shrink the high-n terms.

The same problem happens when computing π using Leibniz's arctangent series (Section 4.2.3)

$$\arctan x = x - \frac{x^3}{3} + \frac{x^5}{5} - \frac{x^7}{7} + \cdots. \qquad (4.36)$$

By using $x = 1$, the direct approximation of $\pi/4$ requires many terms to attain even moderate accuracy. Fortunately, the trigonometric identity $\arctan 1 = 4 \arctan 1/5 - \arctan 1/239$ lowers the largest x to $1/5$ and thereby speeds the convergence.

▶ *Is there an analogous that helps estimate $\ln 2$?*

Because 2 is also $(4/3)/(2/3)$, an analogous rewriting of $\ln 2$ is

$$\ln 2 = \ln \frac{4}{3} - \ln \frac{2}{3}. \qquad (4.37)$$

4.3 Approximating the logarithm

Each fraction has the form $1+x$ with $x = \pm 1/3$. Because x is small, one term of the logarithm series might provide reasonable accuracy. Let's therefore use $\ln(1+x) \approx x$ to approximate the two logarithms:

$$\ln 2 \approx \frac{1}{3} - \left(-\frac{1}{3}\right) = \frac{2}{3}. \tag{4.38}$$

This estimate is accurate to within 5%!

The rewriting trick has helped to compute π (by rewriting the $\arctan x$ series) and to estimate $\ln(1+x)$ (by rewriting x itself). This idea therefore becomes a method—a trick that I use twice (this definition is often attributed to Polya).

> **Problem 4.20   How many terms?**
> The full Taylor series for the logarithm is
>
> $$\ln(1+x) = \sum_{1}^{\infty} (-1)^{n+1} \frac{x^n}{n}. \tag{4.39}$$
>
> If you set $x = 1$ in this series, how many terms are required to estimate $\ln 2$ to within 5%?
>
> **Problem 4.21   Second rewriting**
> Repeat the rewriting method by rewriting 4/3 and 2/3; then estimate $\ln 2$ using only one term of the logarithm series. How accurate is the revised estimate?
>
> **Problem 4.22   Two terms of the Taylor series**
> After rewriting $\ln 2$ as $\ln(4/3) - \ln(2/3)$, use the two-term approximation that $\ln(1+x) \approx x - x^2/2$ to estimate $\ln 2$. Compare the approximation to the one-term estimate, namely 2/3. (Problem 4.24 investigates a pictorial explanation.)
>
> **Problem 4.23   Rational-function approximation for the logarithm**
> The replacement $\ln 2 = \ln(4/3) - \ln(2/3)$ has the general form
>
> $$\ln(1+x) = \ln \frac{1+y}{1-y}, \tag{4.40}$$
>
> where $y = x/(2+x)$.
> Use the expression for y and the one-term series $\ln(1+x) \approx x$ to express $\ln(1+x)$ as a rational function of x (as a ratio of polynomials in x). What are the first few terms of its Taylor series?
>
> Compare those terms to the first few terms of the $\ln(1+x)$ Taylor series, and thereby explain why the rational-function approximation is more accurate than even the two-term series $\ln(1+x) \approx x - x^2/2$.

> **Problem 4.24 Pictorial interpretation of the rewriting**
> a. Use the integral representation of $\ln(1+x)$ to explain why the shaded area is $\ln 2$.
> b. Outline the region that represents
> $$\ln\frac{4}{3} - \ln\frac{2}{3} \qquad (4.41)$$
> when using the circumscribed-rectangle approximation for each logarithm.
> c. Outline the same region when using the trapezoid approximation $\ln(1+x) = x - x^2/2$. Show pictorially that this region, although a different shape, has the same area as the region that you drew in item b.

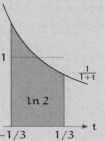

4.4 Bisecting a triangle

Pictorial solutions are especially likely for a geometric problem:

▷ *What is the shortest path that bisects an equilateral triangle into two regions of equal area?*

The possible bisecting paths form an uncountably infinite set. To manage the complexity, try easy cases (Chapter 2)—draw a few equilateral triangles and bisect them with easy paths. Patterns, ideas, or even a solution might emerge.

▷ *What are a few easy paths?*

The simplest bisecting path is a vertical segment that splits the triangle into two right triangles each with base $1/2$. This path is the triangle's altitude, and it has length

$$l = \sqrt{1^2 - (1/2)^2} = \frac{\sqrt{3}}{2} \approx 0.866. \qquad (4.42)$$

An alternative straight path splits the triangle into a trapezoid and a small triangle.

▷ *What is the shape of the smaller triangle, and how long is the path?*

The triangle is similar to the original triangle, so it too is equilateral. Furthermore, it has one-half of the area of the original triangle, so its three

4.4 Bisecting a triangle

sides, one of which is the bisecting path, are a factor of $\sqrt{2}$ smaller than the sides of the original triangle. Thus this path has length $1/\sqrt{2} \approx 0.707$—a substantial improvement on the vertical path with length $\sqrt{3}/2$.

> **Problem 4.25 All one-segment paths**
> An equilateral triangle has infinitely many one-segment bisecting paths. A few of them are shown in the figure. Which one-segment path is the shortest?

Now let's investigate easy two-segment paths. One possible path encloses a diamond and excludes two small triangles. The two small triangles occupy one-half of the entire area. Each small triangle therefore occupies one-fourth of the entire area and has side length $1/2$. Because the bisecting path contains two of these sides, it has length 1. This path is, unfortunately, longer than our two one-segment candidates, whose lengths are $1/\sqrt{2}$ and $\sqrt{3}/2$.

Therefore, a reasonable conjecture is that the shortest path has the fewest segments. This conjecture deserves to be tested (Problem 4.26).

> **Problem 4.26 All two-segment paths**
> Draw a figure showing the variety of two-segment paths. Find the shortest path, showing that it has length
> $$l = 2 \times 3^{1/4} \times \sin 15° \approx 0.681. \tag{4.43}$$
>
> **Problem 4.27 Bisecting with closed paths**
> The bisecting path need not begin or end at an edge of the triangle. Two examples are illustrated here:
>
>
>
> Do you expect closed bisecting paths to be longer or shorter than the shortest one-segment path? Give a geometric reason for your conjecture, and check the conjecture by finding the lengths of the two illustrative closed paths.

▶ *Does using fewer segments produce shorter paths?*

The shortest one-segment path has an approximate length of 0.707; but the shortest two-segment path has an approximate length of 0.681. The length decrease suggests trying extreme paths: paths with an infinite number of segments. In other words, try curved paths. The easiest curved path is probably a circle or a piece of a circle.

▶ *What is a likely candidate for the shortest circle or piece of a circle that bisects the triangle?*

Whether the path is a circle or piece of a circle, it needs a center. However, putting the center inside the triangle and using a full circle produces a long bisecting path (Problem 4.27). The only other plausible center is a vertex of the triangle, so imagine a bisecting arc centered on one vertex.

▶ *How long is this arc?*

The arc subtends one-sixth (60°) of the full circle, so its length is $l = \pi r/3$, where r is radius of the full circle. To find the radius, use the requirement that the arc must bisect the triangle. Therefore, the arc encloses one-half of the triangle's area. The condition on r is that $\pi r^2 = 3\sqrt{3}/4$:

$$\frac{1}{6} \times \underbrace{\text{area of the full circle}}_{\pi r^2} = \frac{1}{2} \times \underbrace{\text{area of the triangle}}_{\sqrt{3}/4}. \tag{4.44}$$

The radius is therefore $(3\sqrt{3}/4\pi)^{1/2}$; the length of the arc is $\pi r/3$, which is approximately 0.673. This curved path is shorter than the shortest two-segment path. It might be the shortest possible path.

To test this conjecture, we use symmetry. Because an equilateral triangle is one-sixth of a hexagon, build a hexagon by replicating the bisected equilateral triangle. Here is the hexagon built from the triangle bisected by a horizontal line:

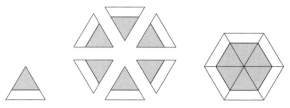

The six bisecting paths form an internal hexagon whose area is one-half of the area of the large hexagon.

▶ *What happens when replicating the triangle bisected by the circular arc?*

4.5 Summing series

When that triangle is replicated, its six copies make a circle with area equal to one-half of the area of the hexagon. For a fixed area, a circle has the shortest perimeter (the isoperimetric theorem [30] and Problem 4.11); therefore, one-sixth of the circle is the shortest bisecting path.

> **Problem 4.28   Replicating the vertical bisection**
> The triangle bisected by a vertical line, if replicated and only rotated, produces a fragmented enclosed region rather than a convex polygon. How can the triangle be replicated so that the six bisecting paths form a regular polygon?
>
> **Problem 4.29   Bisecting the cube**
> Of all surfaces that bisect a cube into two equal volumes, which surface has the smallest area?

4.5 Summing series

For the final example of what pictures can explain, return to the factorial function. Our first approximation to n! began with its integral representation and then used lumping (Section 3.2.3).

Lumping, by replacing a curve with a rectangle whose area is easily computed, is already a pictorial analysis. A second picture for n! begins with the summation representation

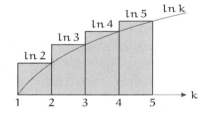

$$\ln n! = \sum_{1}^{n} \ln k. \qquad (4.45)$$

This sum equals the combined area of the circumscribing rectangles.

> **Problem 4.30   Drawing the smooth curve**
> Setting the height of the rectangles requires drawing the $\ln k$ curve—which could intersect the top edge of each rectangle anywhere along the edge. In the preceding figure and the analysis of this section, the curve intersects at the right endpoint of the edge. After reading the section, redo the analysis for two other cases:
>
> a. The curve intersects at the left endpoint of the edge.
> b. The curve intersects at the midpoint of the edge.

That combined area is approximately the area under the $\ln k$ curve, so

$$\ln n! \approx \int_1^n \ln k \, dk = n \ln n - n + 1. \qquad (4.46)$$

Each term in this $\ln n!$ approximation contributes one factor to $n!$:

$$n! \approx n^n \times e^{-n} \times e. \qquad (4.47)$$

Each factor has a counterpart in a factor from Stirling's approximation (Section 3.2.3). In descending order of importance, the factors in Stirling's approximation are

$$n! \approx n^n \times e^{-n} \times \sqrt{n} \times \sqrt{2\pi}. \qquad (4.48)$$

The integral approximation reproduces the two most important factors and almost reproduces the fourth factor: e and $\sqrt{2\pi}$ differ by only 8%. The only unexplained factor is $\sqrt{n}$.

▶ *From where does the $\sqrt{n}$ factor come?*

The $\sqrt{n}$ factor must come from the fragments above the $\ln k$ curve. They are almost triangles and would be easier to add if they were triangles. Therefore, redraw the $\ln k$ curve using straight-line segments (another use of lumping).

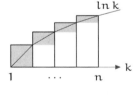

The resulting triangles would be easier to add if they were rectangles. Therefore, let's double each triangle to make it a rectangle.

▶ *What is the sum of these rectangular pieces?*

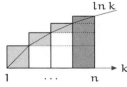

To sum these pieces, lay your right hand along the $k = n$ vertical line. With your left hand, shove the pieces to the right until they hit your right hand. The pieces then stack to form the $\ln n$ rectangle. Because each piece is double the corresponding triangular protrusion, the triangular protrusions sum to $(\ln n)/2$. This triangle correction improves the integral approximation. The resulting approximation for $\ln n!$ now has one more term:

$$\ln n! \approx \underbrace{n \ln n - n + 1}_{\text{integral}} + \underbrace{\frac{\ln n}{2}}_{\text{triangles}}. \qquad (4.49)$$

Upon exponentiating to get $n!$, the correction contributes a factor of $\sqrt{n}$.

$$n! \approx n^n \times e^{-n} \times e \times \sqrt{n}. \qquad (4.50)$$

Compared to Stirling's approximation, the only remaining difference is the factor of e that should be $\sqrt{2\pi}$, an error of only 8%—all from doing one integral and drawing a few pictures.

> **Problem 4.31   Underestimate or overestimate?**
> Does the integral approximation with the triangle correction underestimate or overestimate $n!$? Use pictorial reasoning; then check the conclusion numerically.
>
> **Problem 4.32   Next correction**
> The triangle correction is the first of an infinite series of corrections. The corrections include terms proportional to n^{-2}, n^{-3}, ..., and they are difficult to derive using only pictures. But the n^{-1} correction can be derived with pictures.
>
> a. Draw the regions showing the error made by replacing the smooth $\ln k$ curve with a piecewise-linear curve (a curve made of straight segments).
>
> b. Each region is bounded above by a curve that is almost a parabola, whose area is given by Archimedes' formula (Problem 4.34)
>
> $$\text{area} = \frac{2}{3} \times \text{area of the circumscribing rectangle}. \qquad (4.51)$$
>
> Use that property to approximate the area of each region.
>
> c. Show that when evaluating $\ln n! = \sum_1^n \ln k$, these regions sum to approximately $(1 - n^{-1})/12$.
>
> d. What is the resulting, improved constant term (formerly e) in the approximation to $n!$ and how close is it to $\sqrt{2\pi}$? What factor does the n^{-1} term in the $\ln n!$ approximation contribute to the $n!$ approximation?
>
> These and subsequent corrections are derived in Section 6.3.2 using the technique of analogy.

4.6 Summary and further problems

For tens of millions of years, evolution has refined our perceptual abilities. A small child recognizes patterns more reliably and quickly than does the largest supercomputer. Pictorial reasoning, therefore, taps the mind's vast computational power. It makes us more intelligent by helping us understand and see large ideas at a glance.

For extensive and enjoyable collections of picture proofs, see the works of Nelsen [31, 32]. Here are further problems to develop pictorial reasoning.

Problem 4.33   Another picture for the AM–GM inequality
Sketch $y = \ln x$ to show that the arithmetic mean of a and b is always greater than or equal to their geometric mean, with equality when $a = b$.

Problem 4.34   Archimedes' formula for the area of a parabola
Archimedes showed (long before calculus!) that the closed parabola encloses two-thirds of its circumscribing rectangle. Prove this result by integration.

Show that the closed parabola also encloses two-thirds of the circumscribing parallelogram with vertical sides. These pictorial recipes are useful when approximating functions (for example, in Problem 4.32).

Problem 4.35   Ancient picture for the area of a circle
The ancient Greeks knew that the circumference of a circle with radius r was $2\pi r$. They then used the following picture to show that its area is πr^2. Can you reconstruct the argument?

Problem 4.36   Volume of a sphere
Extend the argument of Problem 4.35 to find the volume of a sphere of radius r, given that its surface area is $4\pi r^2$. Illustrate the argument with a sketch.

Problem 4.37   A famous sum
Use pictorial reasoning to approximate the famous Basel sum $\sum_{1}^{\infty} n^{-2}$.

Problem 4.38   Newton–Raphson method
In general, solving $f(t) = 0$ requires approximations. One method is to start with a guess t_0 and to improve it iteratively using the Newton–Raphson method

$$t_{n+1} = t_n - \frac{f(t_n)}{f'(t_n)}, \qquad (4.52)$$

where $f'(t_n)$ is the derivative df/dt evaluated at $t = t_n$. Draw a picture to justify this recipe; then use the recipe to estimate $\sqrt{2}$. (Then try Problem 4.17.)

5
Taking out the big part

5.1 Multiplication using one and few	77
5.2 Fractional changes and low-entropy expressions	79
5.3 Fractional changes with general exponents	84
5.4 Successive approximation: How deep is the well?	91
5.5 Daunting trigonometric integral	94
5.6 *Summary and further problems*	97

In almost every quantitative problem, the analysis simplifies when you follow the proverbial advice of doing first things first. First approximate and understand the most important effect—the big part—then refine your analysis and understanding. This procedure of successive approximation or "taking out the big part" generates meaningful, memorable, and usable expressions. The following examples introduce the related idea of low-entropy expressions (Section 5.2) and analyze mental multiplication (Section 5.1), exponentiation (Section 5.3), quadratic equations (Section 5.4), and a difficult trigonometric integral (Section 5.5).

5.1 Multiplication using one and few

The first illustration is a method of mental multiplication suited to rough, back-of-the-envelope estimates. The particular calculation is the storage capacity of a data CD-ROM. A data CD-ROM has the same format and storage capacity as a music CD, whose capacity can be estimated as the product of three factors:

$$\underbrace{1\,\text{hr} \times \frac{3600\,\text{s}}{1\,\text{hr}}}_{\text{playing time}} \times \underbrace{\frac{4.4 \times 10^4\,\text{samples}}{1\,\text{s}}}_{\text{sample rate}} \times \underbrace{2\,\text{channels} \times \frac{16\,\text{bits}}{1\,\text{sample}}}_{\text{sample size}}. \qquad (5.1)$$

(In the sample-size factor, the two channels are for stereophonic sound.)

> **Problem 5.1 Sample rate**
> Look up the Shannon–Nyquist sampling theorem [22], and explain why the sample rate (the rate at which the sound pressure is measured) is roughly 40 kHz.
>
> **Problem 5.2 Bits per sample**
> Because $2^{16} \sim 10^5$, a 16-bit sample—as chosen for the CD format—requires electronics accurate to roughly 0.001%. Why didn't the designers of the CD format choose a much larger sample size, say 32 bits (per channel)?
>
> **Problem 5.3 Checking units**
> Check that all the units in the estimate divide out—except for the desired units of bits.

Back-of-the-envelope calculations use rough estimates such as the playing time and neglect important factors such as the bits devoted to error detection and correction. In this and many other estimates, multiplication with 3 decimal places of accuracy would be overkill. An approximate analysis needs an approximate method of calculation.

▶ *What is the data capacity to within a factor of 2?*

The units (the biggest part!) are bits (Problem 5.3), and the three numerical factors contribute $3600 \times 4.4 \times 10^4 \times 32$. To estimate the product, split it into a big part and a correction.

The big part: The most important factor in a back-of-the-envelope product usually comes from the powers of 10, so evaluate this big part first: 3600 contributes three powers of 10, 4.4×10^4 contributes four, and 32 contributes one. The eight powers of 10 produce a factor of 10^8.

The correction: After taking out the big part, the remaining part is a correction factor of $3.6 \times 4.4 \times 3.2$. This product too is simplified by taking out its big part. Round each factor to the closest number among three choices: 1, few, or 10. The invented number few lies midway between 1 and 10: It is the geometric mean of 1 and 10, so $(few)^2 = 10$ and $few \approx 3$. In the product $3.6 \times 4.4 \times 3.2$, each factor rounds to few, so $3.6 \times 4.4 \times 3.2 \approx (few)^3$ or roughly 30.

The units, the powers of 10, and the correction factor combine to give

$$\text{capacity} \sim 10^8 \times 30 \text{ bits} = 3 \times 10^9 \text{ bits}. \tag{5.2}$$

This estimate is within a factor of 2 of the exact product (Problem 5.4), which is itself close to the actual capacity of 5.6×10^9 bits.

> **Problem 5.4 Underestimate or overestimate?**
> Does 3×10^9 overestimate or underestimate $3600 \times 4.4 \times 10^4 \times 32$? Check your reasoning by computing the exact product.
>
> **Problem 5.5 More practice**
> Use the one-or-few method of multiplication to perform the following calculations mentally; then compare the approximate and actual products.
>
> a. $161 \times 294 \times 280 \times 438$. The actual product is roughly 5.8×10^9.
>
> b. Earth's surface area $A = 4\pi R^2$, where the radius is $R \sim 6 \times 10^6$ m. The actual surface area is roughly 5.1×10^{14} m^2.

5.2 Fractional changes and low-entropy expressions

Using the one-or-few method for mental multiplication is fast. For example, 3.15×7.21 quickly becomes few $\times 10^1 \sim 30$, which is within 50% of the exact product 22.7115. To get a more accurate estimate, round 3.15 to 3 and 7.21 to 7. Their product 21 is in error by only 8%. To reduce the error further, one could split 3.15×7.21 into a big part and an additive correction. This decomposition produces

$$(3 + 0.15)(7 + 0.21) = \underbrace{3 \times 7}_{\text{big part}} + \underbrace{0.15 \times 7 + 3 \times 0.21 + 0.15 \times 0.21}_{\text{additive correction}}. \quad (5.3)$$

The approach is sound, but the literal application of taking out the big part produces a messy correction that is hard to remember and understand. Slightly modified, however, taking out the big part provides a clean and intuitive correction. As gravy, developing the improved correction introduces two important street-fighting ideas: fractional changes (Section 5.2.1) and low-entropy expressions (Section 5.2.2). The improved correction will then, as a first of many uses, help us estimate the energy saved by highway speed limits (Section 5.2.3).

5.2.1 Fractional changes

The hygienic alternative to an additive correction is to split the product into a big part and a *multiplicative* correction:

$$3.15 \times 7.21 = \underbrace{3 \times 7}_{\text{big part}} \times \underbrace{(1 + 0.05) \times (1 + 0.03)}_{\text{correction factor}}. \tag{5.4}$$

▶ *Can you find a picture for the correction factor?*

The correction factor is the area of a rectangle with width $1 + 0.05$ and height $1 + 0.03$. The rectangle contains one subrectangle for each term in the expansion of $(1 + 0.05) \times (1 + 0.03)$. Their combined area of roughly $1 + 0.05 + 0.03$ represents an 8% fractional increase over the big part. The big part is 21, and 8% of it is 1.68, so $3.15 \times 7.21 = 22.68$, which is within 0.14% of the exact product.

> **Problem 5.6 Picture for the fractional error**
> What is the pictorial explanation for the fractional error of roughly 0.15%?
>
> **Problem 5.7 Try it yourself**
> Estimate 245×42 by rounding each factor to a nearby multiple of 10, and compare this big part with the exact product. Then draw a rectangle for the correction factor, estimate its area, and correct the big part.

5.2.2 Low-entropy expressions

The correction to 3.15×7.21 was complicated as an absolute or additive change but simple as a fractional change. This contrast is general. Using the additive correction, a two-factor product becomes

$$(x + \Delta x)(y + \Delta y) = xy + \underbrace{x\Delta y + y\Delta x + \Delta x \Delta y}_{\text{additive correction}}. \tag{5.5}$$

> **Problem 5.8 Rectangle picture**
> Draw a rectangle representing the expansion
> $$(x + \Delta x)(y + \Delta y) = xy + x\Delta y + y\Delta x + \Delta x \Delta y. \tag{5.6}$$

When the absolute changes Δx and Δy are small ($x \ll \Delta x$ and $y \ll \Delta y$), the correction simplifies to $x\Delta y + y\Delta x$, but even so it is hard to remember because it has many plausible but incorrect alternatives. For example, it could plausibly contain terms such as $\Delta x \Delta y$, $x\Delta x$, or $y\Delta y$. The extent

5.2 Fractional changes and low-entropy expressions

of the plausible alternatives measures the gap between our intuition and reality; the larger the gap, the harder the correct result must work to fill it, and the harder we must work to remember the correct result.

Such gaps are the subject of statistical mechanics and information theory [20, 21], which define the gap as the logarithm of the number of plausible alternatives and call the logarithmic quantity the entropy. The logarithm does not alter the essential point that expressions differ in the number of plausible alternatives and that high-entropy expressions [28]—ones with many plausible alternatives—are hard to remember and understand.

In contrast, a low-entropy expression allows few plausible alternatives, and elicits, "Yes! How could it be otherwise?!" Much mathematical and scientific progress consists of finding ways of thinking that turn high-entropy expressions into easy-to-understand, low-entropy expressions.

▶ *What is a low-entropy expression for the correction to the product xy?*

A multiplicative correction, being dimensionless, automatically has lower entropy than the additive correction: The set of plausible dimensionless expressions is much smaller than the full set of plausible expressions.

The multiplicative correction is $(x + \Delta x)(y + \Delta y)/xy$. As written, this ratio contains gratuitous entropy. It constructs two dimensioned sums $x + \Delta x$ and $y + \Delta y$, multiplies them, and finally divides the product by xy. Although the result is dimensionless, it becomes so only in the last step. A cleaner method is to group related factors by making dimensionless quantities right away:

$$\frac{(x + \Delta x)(y + \Delta y)}{xy} = \frac{x + \Delta x}{x} \frac{y + \Delta y}{y} = \left(1 + \frac{\Delta x}{x}\right)\left(1 + \frac{\Delta y}{y}\right). \quad (5.7)$$

The right side is built only from the fundamental dimensionless quantity 1 and from meaningful dimensionless ratios: $(\Delta x)/x$ is the fractional change in x, and $(\Delta y)/y$ is the fractional change in y.

The gratuitous entropy came from mixing $x + \Delta x$, $y + \Delta y$, x, and y willy nilly, and it was removed by regrouping or unmixing. Unmixing is difficult with physical systems. Try, for example, to remove a drop of food coloring mixed into a glass of water. The problem is that a glass of water contains roughly 10^{25} molecules. Fortunately, most mathematical expressions have fewer constituents. We can often regroup and unmix the mingled pieces and thereby reduce the entropy of the expression.

> **Problem 5.9   Rectangle for the correction factor**
> Draw a rectangle representing the low-entropy correction factor
> $$\left(1 + \frac{\Delta x}{x}\right)\left(1 + \frac{\Delta y}{y}\right). \tag{5.8}$$

A low-entropy correction factor produces a low-entropy fractional change:

$$\frac{\Delta(xy)}{xy} = \left(1 + \frac{\Delta x}{x}\right)\left(1 + \frac{\Delta y}{y}\right) - 1 = \frac{\Delta x}{x} + \frac{\Delta y}{y} + \frac{\Delta x}{x}\frac{\Delta y}{y}, \tag{5.9}$$

where $\Delta(xy)/xy$ is the fractional change from xy to $(x + \Delta x)(y + \Delta y)$. The rightmost term is the product of two small fractions, so it is small compared to the preceding two terms. Without this small, quadratic term,

$$\frac{\Delta(xy)}{xy} \approx \frac{\Delta x}{x} + \frac{\Delta y}{y}. \tag{5.10}$$

Small fractional changes simply add!

This fractional-change rule is far simpler than the corresponding approximate rule that the absolute change is $x\Delta y + y\Delta x$. Simplicity indicates low entropy; indeed, the only plausible alternative to the proposed rule is the possibility that fractional changes multiply. And this conjecture is not likely: When $\Delta y = 0$, it predicts that $\Delta(xy) = 0$ no matter the value of Δx (this prediction is explored also in Problem 5.12).

> **Problem 5.10   Thermal expansion**
> If, due to thermal expansion, a metal sheet expands in each dimension by 4%, what happens to its area?

> **Problem 5.11   Price rise with a discount**
> Imagine that inflation, or copyright law, increases the price of a book by 10% compared to last year. Fortunately, as a frequent book buyer, you start getting a store discount of 15%. What is the net price change that you see?

5.2.3   Squaring

In analyzing the engineered and natural worlds, a common operation is squaring—a special case of multiplication. Squared lengths are areas, and squared speeds are proportional to the drag on most objects (Section 2.4):

$$F_d \sim \rho v^2 A, \tag{5.11}$$

5.2 Fractional changes and low-entropy expressions

where v is the speed of the object, A is its cross-sectional area, and ρ is the density of the fluid. As a consequence, driving at highway speeds for a distance d consumes an energy $E = F_d d \sim \rho A v^2 d$. Energy consumption can therefore be reduced by driving more slowly. This possibility became important to Western countries in the 1970s when oil prices rose rapidly (see [7] for an analysis). As a result, the United States instituted a highway speed limit of 55 mph (90 kph).

▶ *By what fraction does gasoline consumption fall due to driving 55 mph instead of 65 mph?*

A lower speed limit reduces gasoline consumption by reducing the drag force $\rho A v^2$ and by reducing the driving distance d: People measure and regulate their commuting more by time than by distance. But finding a new home or job is a slow process. Therefore, analyze first things first—assume for this initial analysis that the driving distance d stays fixed (then try Problem 5.14).

With that assumption, E is proportional to v^2, and

$$\frac{\Delta E}{E} = 2 \times \frac{\Delta v}{v}. \tag{5.12}$$

Going from 65 mph to 55 mph is roughly a 15% drop in v, so the energy consumption drops by roughly 30%. Highway driving uses a significant fraction of the oil consumed by motor vehicles, which in the United States consume a significant fraction of all oil consumed. Thus the 30% drop substantially reduced total US oil consumption.

Problem 5.12   A tempting error
If A and x are related by $A = x^2$, a tempting conjecture is that

$$\frac{\Delta A}{A} \approx \left(\frac{\Delta x}{x}\right)^2. \tag{5.13}$$

Disprove this conjecture using easy cases (Chapter 2).

Problem 5.13   Numerical estimates
Use fractional changes to estimate 6.3^3. How accurate is the estimate?

Problem 5.14   Time limit on commuting
Assume that driving time, rather than distance, stays fixed as highway driving speeds fall by 15%. What is the resulting fractional change in the gasoline consumed by highway driving?

> **Problem 5.15   Wind power**
> The power generated by an ideal wind turbine is proportional to v^3 (why?). If wind speeds increase by a mere 10%, what is the effect on the generated power? The quest for fast winds is one reason that wind turbines are placed on cliffs or hilltops or at sea.

5.3 Fractional changes with general exponents

The fractional-change approximations for changes in x^2 (Section 5.2.3) and in x^3 (Problem 5.13) are special cases of the approximation for x^n

$$\frac{\Delta(x^n)}{x^n} \approx n \times \frac{\Delta x}{x}. \tag{5.14}$$

This rule offers a method for mental division (Section 5.3.1), for estimating square roots (Section 5.3.2), and for judging a common explanation for the seasons (Section 5.3.3). The rule requires only that the fractional change be small and that the exponent n not be too large (Section 5.3.4).

5.3.1 Rapid mental division

The special case $n = -1$ provides the method for rapid mental division. As an example, let's estimate $1/13$. Rewrite it as $(x + \Delta x)^{-1}$ with $x = 10$ and $\Delta x = 3$. The big part is $x^{-1} = 0.1$. Because $(\Delta x)/x = 30\%$, the fractional correction to x^{-1} is roughly -30%. The result is 0.07.

$$\frac{1}{13} \approx \frac{1}{10} - 30\% = 0.07, \tag{5.15}$$

where the "-30%" notation, meaning "decrease the previous object by 30%," is a useful shorthand for a factor of $1 - 0.3$.

▶ *How accurate is the estimate, and what is the source of the error?*

The estimate is in error by only 9%. The error arises because the linear approximation

$$\frac{\Delta(x^{-1})}{x^{-1}} \approx -1 \times \frac{\Delta x}{x} \tag{5.16}$$

does not include the square (or higher powers) of the fractional change $(\Delta x)/x$ (Problem 5.17 asks you to find the squared term).

5.3 Fractional changes with general exponents

▶ *How can the error in the linear approximation be reduced?*

To reduce the error, reduce the fractional change. Because the fractional change is determined by the big part, let's increase the accuracy of the big part. Accordingly, multiply 1/13 by 8/8, a convenient form of 1, to construct 8/104. Its big part 0.08 approximates 1/13 already to within 4%. To improve it, write 1/104 as $(x + \Delta x)^{-1}$ with $x = 100$ and $\Delta x = 4$. The fractional change $(\Delta x)/x$ is now 0.04 (rather than 0.3); and the fractional correction to $1/x$ and $8/x$ is a mere -4%. The corrected estimate is 0.0768:

$$\frac{1}{13} \approx 0.08 - 4\% = 0.08 - 0.0032 = 0.0768. \tag{5.17}$$

This estimate can be done mentally in seconds and is accurate to 0.13%!

> **Problem 5.16 Next approximation**
> Multiply 1/13 by a convenient form of 1 to make a denominator near 1000; then estimate 1/13. How accurate is the resulting approximation?
>
> **Problem 5.17 Quadratic approximation**
> Find A, the coefficient of the quadratic term in the improved fractional-change approximation
>
> $$\frac{\Delta(x^{-1})}{x^{-1}} \approx -1 \times \frac{\Delta x}{x} + A \times \left(\frac{\Delta x}{x}\right)^2. \tag{5.18}$$
>
> Use the resulting approximation to improve the estimates for 1/13.
>
> **Problem 5.18 Fuel efficiency**
> Fuel efficiency is inversely proportional to energy consumption. If a 55 mph speed limit decreases energy consumption by 30%, what is the new fuel efficiency of a car that formerly got 30 miles per US gallon (12.8 kilometers per liter)?

5.3.2 Square roots

The fractional exponent $n = 1/2$ provides the method for estimating square roots. As an example, let's estimate $\sqrt{10}$. Rewrite it as $(x + \Delta x)^{1/2}$ with $x = 9$ and $\Delta x = 1$. The big part $x^{1/2}$ is 3. Because $(\Delta x)/x = 1/9$ and $n = 1/2$, the fractional correction is 1/18. The corrected estimate is

$$\sqrt{10} \approx 3 \times \left(1 + \frac{1}{18}\right) \approx 3.1667. \tag{5.19}$$

The exact value is 3.1622..., so the estimate is accurate to 0.14%.

> **Problem 5.19 Overestimate or underestimate?**
> Does the linear fractional-change approximation overestimate all square roots (as it overestimated $\sqrt{10}$)? If yes, explain why; if no, give a counterexample.
>
> **Problem 5.20 Cosine approximation**
> Use the small-angle approximation $\sin\theta \approx \theta$ to show that $\cos\theta \approx 1 - \theta^2/2$.
>
> **Problem 5.21 Reducing the fractional change**
> To reduce the fractional change when estimating $\sqrt{10}$, rewrite it as $\sqrt{360}/6$ and then estimate $\sqrt{360}$. How accurate is the resulting estimate for $\sqrt{10}$?
>
> **Problem 5.22 Another method to reduce the fractional change**
> Because $\sqrt{2}$ is fractionally distant from the nearest integer square roots $\sqrt{1}$ and $\sqrt{4}$, fractional changes do not give a direct and accurate estimate of $\sqrt{2}$. A similar problem occurred in estimating $\ln 2$ (Section 4.3); there, rewriting 2 as $(4/3)/(2/3)$ improved the accuracy. Does that rewriting help estimate $\sqrt{2}$?
>
> **Problem 5.23 Cube root**
> Estimate $2^{1/3}$ to within 10%.

5.3.3 A reason for the seasons?

Summers are warmer than winters, it is often alleged, because the earth is closer to the sun in the summer than in the winter. This common explanation is bogus for two reasons. First, summers in the southern hemisphere happen alongside winters in the northern hemisphere, despite almost no difference in the respective distances to the sun. Second, as we will now estimate, the varying earth–sun distance produces too small a temperature difference. The causal chain—that the distance determines the intensity of solar radiation and that the intensity determines the surface temperature—is most easily analyzed using fractional changes.

Intensity of solar radiation: The intensity is the solar power divided by the area over which it spreads. The solar power hardly changes over a year (the sun has existed for several billion years); however, at a distance r from the sun, the energy has spread over a giant sphere with surface area $\sim r^2$. The intensity I therefore varies according to $I \propto r^{-2}$. The fractional changes in radius and intensity are related by

$$\frac{\Delta I}{I} \approx -2 \times \frac{\Delta r}{r}. \tag{5.20}$$

5.3 Fractional changes with general exponents

Surface temperature: The incoming solar energy cannot accumulate and returns to space as blackbody radiation. Its outgoing intensity depends on the earth's surface temperature T according to the Stefan–Boltzmann law $I = \sigma T^4$ (Problem 1.12), where σ is the Stefan–Boltzmann constant. Therefore $T \propto I^{1/4}$. Using fractional changes,

$$\frac{\Delta T}{T} \approx \frac{1}{4} \times \frac{\Delta I}{I}. \tag{5.21}$$

This relation connects intensity and temperature. The temperature and distance are connected by $(\Delta I)/I = -2 \times (\Delta r)/r$. When joined, the two relations connect distance and temperature as follows:

$$\frac{\Delta r}{r} \xrightarrow{\boxed{-2}} \underset{I \propto r^{-2}}{\frac{\Delta I}{I} \approx -2 \times \frac{\Delta r}{r}} \xrightarrow{\boxed{\tfrac{1}{4}}} \underset{T \propto I^{1/4}}{\frac{\Delta T}{T} \approx -\frac{1}{2} \times \frac{\Delta r}{r}}$$

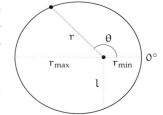

The next step in the computation is to estimate the input $(\Delta r)/r$—namely, the fractional change in the earth–sun distance. The earth orbits the sun in an ellipse; its orbital distance is

$$r = \frac{l}{1 + \epsilon \cos \theta}, \tag{5.22}$$

where ϵ is the eccentricity of the orbit, θ is the polar angle, and l is the semilatus rectum. Thus r varies from $r_{min} = l/(1+\epsilon)$ (when $\theta = 0°$) to $r_{max} = l/(1-\epsilon)$ (when $\theta = 180°$). The increase from r_{min} to l contributes a fractional change of roughly ϵ. The increase from l to r_{max} contributes another fractional change of roughly ϵ. Thus, r varies by roughly 2ϵ. For the earth's orbit, $\epsilon = 0.016$, so the earth–sun distance varies by 0.032 or 3.2% (making the intensity vary by 6.4%).

Problem 5.24   Where is the sun?
The preceding diagram of the earth's orbit placed the sun away from the center of the ellipse. The diagram to the right shows the sun at an alternative and perhaps more natural location: at the center of the ellipse. What physical laws, if any, prevent the sun from sitting at the center of the ellipse?

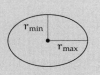

Problem 5.25   Check the fractional change
Look up the minimum and maximum earth–sun distances and check that the distance does vary by 3.2% from minimum to maximum.

A 3.2% increase in distance causes a slight drop in temperature:

$$\frac{\Delta T}{T} \approx -\frac{1}{2} \times \frac{\Delta r}{r} = -1.6\%. \tag{5.23}$$

However, man does not live by fractional changes alone and experiences the absolute temperature change ΔT.

$$\Delta T = -1.6\% \times T. \tag{5.24}$$

▶ *In winter* $T \approx 0°\,C$, *so is* $\Delta T \approx 0°\,C$?

If our calculation predicts that $\Delta T \approx 0°\,C$, it must be wrong. An even less plausible conclusion results from measuring T in Fahrenheit degrees, which makes T often negative in parts of the northern hemisphere. Yet ΔT cannot flip its sign just because T is measured in Fahrenheit degrees!

Fortunately, the temperature scale is constrained by the Stefan–Boltzmann law. For blackbody flux to be proportional to T^4, temperature must be measured relative to a state with zero thermal energy: absolute zero. Neither the Celsius nor the Fahrenheit scale satisfies this requirement.

In contrast, the Kelvin scale does measure temperature relative to absolute zero. On the Kelvin scale, the average surface temperature is $T \approx 300\,K$; thus, a 1.6% change in T makes $\Delta T \approx 5\,K$. A $5\,K$ change is also a $5°\,C$ change—Kelvin and Celsius degrees are the same size, although the scales have different zero points. (See also Problem 5.26.) A typical temperature change between summer and winter in temperate latitudes is $20°\,C$—much larger than the predicted $5°\,C$ change, even after allowing for errors in the estimate. A varying earth–sun distance is a dubious explanation of the reason for the seasons.

> **Problem 5.26   Converting to Fahrenheit**
> The conversion between Fahrenheit and Celsius temperatures is
>
> $$F = 1.8C + 32, \tag{5.25}$$
>
> so a change of $5°\,C$ should be a change of $41°\,F$—sufficiently large to explain the seasons! What is wrong with this reasoning?
>
> **Problem 5.27   Alternative explanation**
> If a varying distance to the sun cannot explain the seasons, what can? Your proposal should, in passing, explain why the northern and southern hemispheres have summer 6 months apart.

5.3 Fractional changes with general exponents

5.3.4 Limits of validity

The linear fractional-change approximation

$$\frac{\Delta(x^n)}{x^n} \approx n \times \frac{\Delta x}{x} \tag{5.26}$$

has been useful. But when is it valid? To investigate without drowning in notation, write z for Δx; then choose $x = 1$ to make z the absolute and the fractional change. The right side becomes nz, and the linear fractional-change approximation is equivalent to

$$(1+z)^n \approx 1 + nz. \tag{5.27}$$

The approximation becomes inaccurate when z is too large: for example, when evaluating $\sqrt{1+z}$ with $z = 1$ (Problem 5.22). Is the exponent n also restricted? The preceding examples illustrated only moderate-sized exponents: $n = 2$ for energy consumption (Section 5.2.3), -2 for fuel efficiency (Problem 5.18), -1 for reciprocals (Section 5.3.1), $1/2$ for square roots (Section 5.3.2), and -2 and $1/4$ for the seasons (Section 5.3.3). We need further data.

▶ *What happens in the extreme case of large exponents?*

With a large exponent such as $n = 100$ and, say, $z = 0.001$, the approximation predicts that $1.001^{100} \approx 1.1$—close to the true value of $1.105\ldots$ However, choosing the same n alongside $z = 0.1$ (larger than 0.001 but still small) produces the terrible prediction

$$\underbrace{1.1^{100}}_{(1+z)^n} = 1 + \underbrace{100 \times 0.1}_{nz} = 11; \tag{5.28}$$

1.1^{100} is roughly $14{,}000$, more than 1000 times larger than the prediction.

Both predictions used large n and small z, yet only one prediction was accurate; thus, the problem cannot lie in n or z alone. Perhaps the culprit is the dimensionless product nz. To test that idea, hold nz constant while trying large values of n. For nz, a sensible constant is 1—the simplest dimensionless number. Here are several examples.

$$\begin{aligned} 1.1^{10} &\approx 2.59374, \\ 1.01^{100} &\approx 2.70481, \\ 1.001^{1000} &\approx 2.71692. \end{aligned} \tag{5.29}$$

In each example, the approximation incorrectly predicts that $(1+z)^n = 2$.

▶ *What is the cause of the error?*

To find the cause, continue the sequence beyond 1.001^{1000} and hope that a pattern will emerge: The values seem to approach $e = 2.718281828\ldots$, the base of the natural logarithms. Therefore, take the logarithm of the whole approximation.

k	$(1+10^{-k})^{10^k}$
1	2.5937425
2	2.7048138
3	2.7169239
4	2.7181459
5	2.7182682
6	2.7182805
7	2.7182817

$$\ln(1+z)^n = n\ln(1+z). \tag{5.30}$$

Pictorial reasoning showed that $\ln(1+z) \approx z$ when $z \ll 1$ (Section 4.3). Thus, $n\ln(1+z) \approx nz$, making $(1+z)^n \approx e^{nz}$. This improved approximation explains why the approximation $(1+z)^n \approx 1+nz$ failed with large nz: Only when $nz \ll 1$ is e^{nz} approximately $1+nz$. Therefore, when $z \ll 1$ the two simplest approximation are

$$(1+z)^n \approx \begin{cases} 1+nz & (z \ll 1 \text{ and } nz \ll 1), \\ e^{nz} & (z \ll 1 \text{ and } nz \text{ unrestricted}). \end{cases} \tag{5.31}$$

The diagram shows, across the whole n–z plane, the simplest approximation in each region. The axes are logarithmic and n and z are assumed positive: The right half plane shows $z \gg 1$, and the upper half plane shows $n \gg 1$. On the lower right, the boundary curve is $n\ln z = 1$. Explaining the boundaries and extending the approximations is an instructive exercise (Problem 5.28).

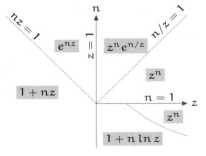

> **Problem 5.28 Explaining the approximation plane**
> In the right half plane, explain the $n/z = 1$ and $n\ln z = 1$ boundaries. For the whole plane, relax the assumption of positive n and z as far as possible.
>
> **Problem 5.29 Binomial-theorem derivation**
> Try the following alternative derivation of $(1+z)^n \approx e^{nz}$ (where $n \gg 1$). Expand $(1+z)^n$ using the binomial theorem, simplify the products in the binomial coefficients by approximating $n-k$ as n, and compare the resulting expansion to the Taylor series for e^{nz}.

5.4 Successive approximation: How deep is the well?

The next illustration of taking out the big part emphasizes successive approximation and is disguised as a physics problem.

> You drop a stone down a well of unknown depth h and hear the splash 4 s later. Neglecting air resistance, find h to within 5%. Use $c_s = 340 \text{ m s}^{-1}$ as the speed of sound and $g = 10 \text{ m s}^{-2}$ as the strength of gravity.

Approximate and exact solutions give almost the same well depth, but offer significantly different understandings.

5.4.1 Exact depth

The depth is determined by the constraint that the 4 s wait splits into two times: the rock falling freely down the well and the sound traveling up the well. The free-fall time is $\sqrt{2h/g}$ (Problem 1.3), so the total time is

$$T = \underbrace{\sqrt{\frac{2h}{g}}}_{\text{rock}} + \underbrace{\frac{h}{c_s}}_{\text{sound}}. \tag{5.32}$$

To solve for h exactly, either isolate the square root on one side and square both sides to get a quadratic equation in h (Problem 5.30); or, for a less error-prone method, rewrite the constraint as a quadratic equation in a new variable $z = \sqrt{h}$.

> **Problem 5.30 Other quadratic**
> Solve for h by isolating the square root on one side and squaring both sides. What are the advantages and disadvantages of this method in comparison with the method of rewriting the constraint as a quadratic in $z = \sqrt{h}$?

As a quadratic equation in $z = \sqrt{h}$, the constraint is

$$\frac{1}{c_s} z^2 + \sqrt{\frac{2}{g}} z - T = 0. \tag{5.33}$$

Using the quadratic formula and choosing the positive root yields

$$z = \frac{-\sqrt{2/g} + \sqrt{2/g + 4T/c_s}}{2/c_s}. \tag{5.34}$$

Because $z^2 = h$,

$$h = \left(\frac{-\sqrt{2/g} + \sqrt{2/g + 4T/c_s}}{2/c_s}\right)^2. \tag{5.35}$$

Substituting $g = 10\,\mathrm{m\,s^{-2}}$ and $c_s = 340\,\mathrm{m\,s^{-1}}$ gives $h \approx 71.56\,\mathrm{m}$.

Even if the depth is correct, the exact formula for it is a mess. Such high-entropy horrors arise frequently from the quadratic formula; its use often signals the triumph of symbol manipulation over thought. Exact answers, we will find, may be less useful than approximate answers.

5.4.2 Approximate depth

To find a low-entropy, approximate depth, identify the big part—the most important effect. Here, most of the total time is the rock's free fall: The rock's maximum speed, even if it fell for the entire 4 s, is only $gT = 40\,\mathrm{m\,s^{-1}}$, which is far below c_s. Therefore, the most important effect should arise in the extreme case of infinite sound speed.

▶ *If $c_s = \infty$, how deep is the well?*

In this zeroth approximation, the free-fall time t_0 is the full time $T = 4\,\mathrm{s}$, so the well depth h_0 becomes

$$h_0 = \frac{1}{2}gt_0^2 = 80\,\mathrm{m}. \tag{5.36}$$

▶ *Is this approximate depth an overestimate or underestimate? How accurate is it?*

This approximation neglects the sound-travel time, so it overestimates the free-fall time and therefore the depth. Compared to the true depth of roughly 71.56 m, it overestimates the depth by only 11%—reasonable accuracy for a quick method offering physical insight. Furthermore, this approximation suggests its own refinement.

▶ *How can this approximation be improved?*

To improve it, use the approximate depth h_0 to approximate the sound-travel time.

$$t_\mathrm{sound} \approx \frac{h_0}{c_s} \approx 0.24\,\mathrm{s}. \tag{5.37}$$

$T \longrightarrow t \xrightarrow{\frac{1}{2}gt^2} h$

$T - \frac{h}{c_s}$

The remaining time is the next approximation to the free-fall time.

5.4 Successive approximation: How deep is the well?

$$t_1 = T - \frac{h_0}{c_s} \approx 3.76\,\text{s}. \tag{5.38}$$

In that time, the rock falls a distance $gt_1^2/2$, so the next approximation to the depth is

$$h_1 = \frac{1}{2}gt_1^2 \approx 70.87\,\text{m}. \tag{5.39}$$

▶ *Is this approximate depth an overestimate or underestimate? How accurate is it?*

The calculation of h_1 used h_0 to estimate the sound-travel time. Because h_0 overestimates the depth, the procedure overestimates the sound-travel time and, by the same amount, underestimates the free-fall time. Thus h_1 underestimates the depth. Indeed, h_1 is slightly smaller than the true depth of roughly 71.56 m—but by only 1.3%.

The method of successive approximation has several advantages over solving the quadratic formula exactly. First, it helps us develop a physical understanding of the system; we realize, for example, that most of the $T = 4\,\text{s}$ is spent in free fall, so the depth is roughly $gT^2/2$. Second, it has a pictorial explanation (Problem 5.34). Third, it gives a sufficiently accurate answer quickly. If you want to know whether it is safe to jump into the well, why calculate the depth to three decimal places?

Finally, the method can handle small changes in the model. Maybe the speed of sound varies with depth, or air resistance becomes important (Problem 5.32). Then the brute-force, quadratic-formula method fails. The quadratic formula and the even messier cubic and the quartic formulas are rare closed-form solutions to complicated equations. Most equations have no closed-form solution. Therefore, a small change to a solvable model usually produces an intractable model—if we demand an exact answer. The method of successive approximation is a robust alternative that produces low-entropy, comprehensible solutions.

> **Problem 5.31 Parameter-value inaccuracies**
> What is h_2, the second approximation to the depth? Compare the error in h_1 and h_2 with the error made by using $g = 10\,\text{m}\,\text{s}^{-2}$.
>
> **Problem 5.32 Effect of air resistance**
> Roughly what fractional error in the depth is produced by neglecting air resistance (Section 2.4.2)? Compare this error to the error in the first approximation h_1 and in the second approximation h_2 (Problem 5.31).

Problem 5.33 Dimensionless form of the well-depth analysis

Even the messiest results are cleaner and have lower entropy in dimensionless form. The four quantities h, g, T, and c_s produce two independent dimensionless groups (Section 2.4.1). An intuitively reasonable pair are

$$\overline{h} \equiv \frac{h}{gT^2} \quad \text{and} \quad \overline{T} \equiv \frac{gT}{c_s}. \qquad (5.40)$$

a. What is a physical interpretation of $\overline{T}$?

b. With two groups, the general dimensionless form is $\overline{h} = f(\overline{T})$. What is $\overline{h}$ in the easy case $\overline{T} \to 0$?

c. Rewrite the quadratic-formula solution

$$h = \left(\frac{-\sqrt{2/g} + \sqrt{2/g + 4T/c_s}}{2/c_s} \right)^2 \qquad (5.41)$$

as $\overline{h} = f(\overline{T})$. Then check that $f(\overline{T})$ behaves correctly in the easy case $\overline{T} \to 0$.

Problem 5.34 Spacetime diagram of the well depth

How does the spacetime diagram [44] illustrate the successive approximation of the well depth? On the diagram, mark h_0 (the zeroth approximation to the depth), h_1, and the exact depth h. Mark t_0, the zeroth approximation to the free-fall time. Why are portions of the rock and sound-wavefront curves dotted? How would you redraw the diagram if the speed of sound doubled? If g doubled?

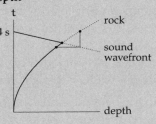

5.5 Daunting trigonometric integral

The final example of taking out the big part is to estimate a daunting trigonometric integral that I learned as an undergraduate. My classmates and I spent many late nights in the physics library solving homework problems; the graduate students, doing the same for their courses, would regale us with their favorite mathematics and physics problems.

The integral appeared on the mathematical-preliminaries exam to enter the Landau Institute for Theoretical Physics in the former USSR. The problem is to evaluate

$$\int_{-\pi/2}^{\pi/2} (\cos t)^{100} \, dt \qquad (5.42)$$

5.5 Daunting trigonometric integral

to within 5% in less than 5 min without using a calculator or computer!

That $(\cos t)^{100}$ looks frightening. Most trigonometric identities do not help. The usually helpful identity $(\cos t)^2 = (\cos 2t - 1)/2$ produces only

$$(\cos t)^{100} = \left(\frac{\cos 2t - 1}{2}\right)^{50}, \tag{5.43}$$

which becomes a trigonometric monster upon expanding the 50th power.

A clue pointing to a simpler method is that 5% accuracy is sufficient—so, find the big part! The integrand is largest when t is near zero. There, $\cos t \approx 1 - t^2/2$ (Problem 5.20), so the integrand is roughly

$$(\cos t)^{100} \approx \left(1 - \frac{t^2}{2}\right)^{100}. \tag{5.44}$$

It has the familiar form $(1 + z)^n$, with fractional change $z = -t^2/2$ and exponent $n = 100$. When t is small, $z = -t^2/2$ is tiny, so $(1 + z)^n$ may be approximated using the results of Section 5.3.4:

$$(1 + z)^n \approx \begin{cases} 1 + nz & (z \ll 1 \text{ and } nz \ll 1) \\ e^{nz} & (z \ll 1 \text{ and } nz \text{ unrestricted}). \end{cases} \tag{5.45}$$

Because the exponent n is large, nz can be large even when t and z are small. Therefore, the safest approximation is $(1 + z)^n \approx e^{nz}$; then

$$(\cos t)^{100} \approx \left(1 - \frac{t^2}{2}\right)^{100} \approx e^{-50t^2}. \tag{5.46}$$

A cosine raised to a high power becomes a Gaussian! As a check on this surprising conclusion, computer-generated plots of $(\cos t)^n$ for $n = 1 \ldots 5$ show a Gaussian bell shape taking form as n increases.

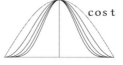

Even with this graphical evidence, replacing $(\cos t)^{100}$ by a Gaussian is a bit suspicious. In the original integral, t ranges from $-\pi/2$ to $\pi/2$, and these endpoints are far outside the region where $\cos t \approx 1 - t^2/2$ is an accurate approximation. Fortunately, this issue contributes only a tiny error (Problem 5.35). Ignoring this error turns the original integral into a Gaussian integral with finite limits:

$$\int_{-\pi/2}^{\pi/2} (\cos t)^{100} \, dt \approx \int_{-\pi/2}^{\pi/2} e^{-50t^2} \, dt. \tag{5.47}$$

Unfortunately, with finite limits the integral has no closed form. But extending the limits to infinity produces a closed form while contributing almost no error (Problem 5.36). The approximation chain is now

$$\int_{-\pi/2}^{\pi/2} (\cos t)^{100} \, dt \approx \int_{-\pi/2}^{\pi/2} e^{-50t^2} \, dt \approx \int_{-\infty}^{\infty} e^{-50t^2} \, dt. \qquad (5.48)$$

> **Problem 5.35 Using the original limits**
> The approximation $\cos t \approx 1 - t^2/2$ requires that t be small. Why doesn't using the approximation outside the small-t range contribute a significant error?
>
> **Problem 5.36 Extending the limits**
> Why doesn't extending the integration limits from $\pm\pi/2$ to $\pm\infty$ contribute a significant error?

The last integral is an old friend (Section 2.1): $\int_{-\infty}^{\infty} e^{-\alpha t^2} \, dt = \sqrt{\pi/\alpha}$. With $\alpha = 50$, the integral becomes $\sqrt{\pi/50}$. Conveniently, 50 is roughly 16π, so the square root—and our 5% estimate—is roughly 0.25.

For comparison, the exact integral is (Problem 5.41)

$$\int_{-\pi/2}^{\pi/2} (\cos t)^n \, dt = 2^{-n} \binom{n}{n/2} \pi. \qquad (5.49)$$

When $n = 100$, the binomial coefficient and power of two produce

$$\frac{126114180681955241668515562157}{1584563250285286751870879006 72} \pi \approx 0.25003696348037. \qquad (5.50)$$

Our 5-minute, within-5% estimate of 0.25 is accurate to almost 0.01%!

> **Problem 5.37 Sketching the approximations**
> Plot $(\cos t)^{100}$ and its two approximations e^{-50t^2} and $1 - 50t^2$.
>
> **Problem 5.38 Simplest approximation**
> Use the linear fractional-change approximation $(1 - t^2/2)^{100} \approx 1 - 50t^2$ to approximate the integrand; then integrate it over the range where $1 - 50t^2$ is positive. How close is the result of this 1-minute method to the exact value $0.2500\ldots$?
>
> **Problem 5.39 Huge exponent**
> Estimate
> $$\int_{-\pi/2}^{\pi/2} (\cos t)^{10000} \, dt. \qquad (5.51)$$

Problem 5.40   How low can you go?
Investigate the accuracy of the approximation

$$\int_{-\pi/2}^{\pi/2} (\cos t)^n \, dt \approx \sqrt{\frac{\pi}{n}}, \qquad (5.52)$$

for small n, including $n = 1$.

Problem 5.41   Closed form
To evaluate the integral

$$\int_{-\pi/2}^{\pi/2} (\cos t)^{100} \, dt \qquad (5.53)$$

in closed form, use the following steps:

a. Replace $\cos t$ with $(e^{it} + e^{-it})/2$.
b. Use the binomial theorem to expand the 100th power.
c. Pair each term like e^{ikt} with a counterpart e^{-ikt}; then integrate their sum from $-\pi/2$ to $\pi/2$. What value or values of k produce a sum whose integral is nonzero?

5.6 Summary and further problems

Upon meeting a complicated problem, divide it into a big part—the most important effect—and a correction. Analyze the big part first, and worry about the correction afterward. This successive-approximation approach, a species of divide-and-conquer reasoning, gives results automatically in a low-entropy form. Low-entropy expressions admit few plausible alternatives; they are therefore memorable and comprehensible. In short, approximate results can be more useful than exact results.

Problem 5.42   Large logarithm
What is the big part in $\ln(1+e^2)$? Give a short calculation to estimate $\ln(1+e^2)$ to within 2%.

Problem 5.43   Bacterial mutations
In an experiment described in a Caltech biology seminar in the 1990s, researchers repeatedly irradiated a population of bacteria in order to generate mutations. In each round of radiation, 5% of the bacteria got mutated. After 140 rounds, roughly what fraction of bacteria were left unmutated? (The seminar speaker gave the audience 3 s to make a guess, hardly enough time to use or even find a calculator.)

Problem 5.44 Quadratic equations revisited
The following quadratic equation, inspired by [29], describes a very strongly damped oscillating system.
$$s^2 + 10^9 s + 1 = 0. \tag{5.54}$$

a. Use the quadratic formula and a standard calculator to find both roots of the quadratic. What goes wrong and why?

b. Estimate the roots by taking out the big part. (*Hint:* Approximate and solve the equation in appropriate extreme cases.) Then improve the estimates using successive approximation.

c. What are the advantages and disadvantages of the quadratic-formula analysis versus successive approximation?

Problem 5.45 Normal approximation to the binomial distribution
The binomial expansion
$$\left(\frac{1}{2} + \frac{1}{2}\right)^{2n} \tag{5.55}$$

contains terms of the form
$$f(k) \equiv \binom{2n}{n-k} 2^{-2n}, \tag{5.56}$$

where $k = -n \ldots n$. Each term $f(k)$ is the probability of tossing $n - k$ heads (and $n + k$ tails) in $2n$ coin flips; $f(k)$ is the so-called binomial distribution with parameters $p = q = 1/2$. Approximate this distribution by answering the following questions:

a. Is $f(k)$ an even or an odd function of k? For what k does $f(k)$ have its maximum?

b. Approximate $f(k)$ when $k \ll n$ and sketch $f(k)$. Therefore, derive and explain the normal approximation to the binomial distribution.

c. Use the normal approximation to show that the variance of this binomial distribution is $n/2$.

Problem 5.46 Beta function
The following integral appears often in Bayesian inference:
$$f(a, b) = \int_0^1 x^a (1-x)^b \, dx, \tag{5.57}$$

where $f(a-1, b-1)$ is the Euler beta function. Use street-fighting methods to conjecture functional forms for $f(a, 0)$, $f(a, a)$, and, finally, $f(a, b)$. Check your conjectures with a high-quality table of integrals or a computer-algebra system such as Maxima.

6
Analogy

6.1 Spatial trigonometry: The bond angle in methane	99
6.2 Topology: How many regions?	103
6.3 Operators: Euler–MacLaurin summation	107
6.4 Tangent roots: A daunting transcendental sum	113
6.5 *Bon voyage*	121

When the going gets tough, the tough lower their standards. This idea, the theme of the whole book, underlies the final street-fighting tool of reasoning by analogy. Its advice is simple: Faced with a difficult problem, construct and solve a similar but simpler problem—an analogous problem. Practice develops fluency. The tool is introduced in spatial trigonometry (Section 6.1); sharpened on solid geometry and topology (Section 6.2); then applied to discrete mathematics (Section 6.3) and, in the farewell example, to an infinite transcendental sum (Section 6.4).

6.1 Spatial trigonometry: The bond angle in methane

The first analogy comes from spatial trigonometry. In methane (chemical formula CH_4), a carbon atom sits at the center of a regular tetrahedron, and one hydrogen atom sits at each vertex. What is the angle θ between two carbon–hydrogen bonds?

Angles in three dimensions are hard to visualize. Try, for example, to imagine and calculate the angle between two faces of a regular tetrahedron. Because two-dimensional angles are easy to visualize, let's construct and analyze an analogous planar molecule. Knowing its bond angle might help us guess methane's bond angle.

▶ *Should the analogous planar molecule have four or three hydrogens?*

Four hydrogens produce four bonds which, when spaced regularly in a plane, produce two different bond angles. In contrast, methane contains only one bond angle. Therefore, using four hydrogens alters a crucial feature of the original problem. The likely solution is to construct the analogous planar molecule using only three hydrogens.

Three hydrogens arranged regularly in a plane create only one bond angle: $\theta = 120°$. Perhaps this angle is the bond angle in methane! One data point, however, is a thin reed on which to hang a prediction for higher dimensions. The single data point for two dimensions ($d = 2$) is consistent with numerous conjectures—for example, that in d dimensions the bond angle is 120° or $(60d)°$ or much else.

Selecting a reasonable conjecture requires gathering further data. Easily available data comes from an even simpler yet analogous problem: the one-dimensional, linear molecule CH_2. Its two hydrogens sit opposite one another, so the two C–H bonds form an angle of $\theta = 180°$.

▶ *Based on the accumulated data, what are reasonable conjectures for the three-dimensional angle θ_3?*

The one-dimensional molecule eliminates the conjecture that $\theta_d = (60d)°$. It also suggests new conjectures—for example, that $\theta_d = (240 - 60d)°$ or $\theta_d = 360°/(d+1)$. Testing these conjectures is an ideal task for the method of easy cases. The easy-cases test of higher dimensions (high d) refutes the conjecture that $\theta_d = (240 - 60d)°$. For high d, it predicts implausible bond angles—namely, $\theta = 0$ for $d = 4$ and $\theta < 0$ for $d > 4$.

d	θ_d
1	180°
2	120
3	?

Fortunately, the second suggestion, $\theta_d = 360°/(d+1)$, passes the same easy-cases test. Let's continue to test it by evaluating its prediction for methane—namely, $\theta_3 = 90°$. Imagine then a big brother of methane: a CH_6 molecule with carbon at the center of a cube and six hydrogens at the face centers. Its small bond angle is 90°. (The other bond angle is 180°.) Now remove two hydrogens to turn CH_6 into CH_4, evenly spreading out the remaining four hydrogens. Reducing the crowding raises the small bond angle above 90°—and refutes the prediction that $\theta_3 = 90°$.

6.1 Spatial trigonometry: The bond angle in methane

> **Problem 6.1 How many hydrogens?**
> How many hydrogens are needed in the analogous four- and five-dimensional bond-angle problems? Use this information to show that $\theta_4 > 90°$. Is $\theta_d > 90°$ for all d?

The data so far have refuted the simplest rational-function conjectures $(240 - 60d)°$ and $360°/(d+1)$. Although other rational-function conjectures might survive, with only two data points the possibilities are too vast. Worse, θ_d might not even be a rational function of d.

Progress requires a new idea: The bond angle might not be the simplest variable to study. An analogous difficulty arises when conjecturing the next term in the series 3, 5, 11, 29, ...

▷ *What is the next term in the series?*

At first glance, the numbers seems almost random. Yet subtracting 2 from each term produces 1, 3, 9, 27, ... Thus, in the original series the next term is likely to be 83. Similarly, a simple transformation of the θ_d data might help us conjecture a pattern for θ_d.

▷ *What transformation of the θ_d data produces simple patterns?*

The desired transformation should produce simple patterns and have aesthetic or logical justification. One justification is the structure of an honest calculation of the bond angle, which can be computed as a dot product of two C–H vectors (Problem 6.3). Because dot products involve cosines, a worthwhile transformation of θ_d is $\cos \theta_d$.

This transformation simplifies the data: The $\cos \theta_d$ series begins simply $-1, -1/2, \ldots$ Two plausible continuations are $-1/4$ or $-1/3$; they correspond, respectively, to the general term $-1/2^{d-1}$ or $-1/d$.

d	θ_d	$\cos \theta_d$
1	180°	-1
2	120	$-1/2$
3	?	?

▷ *Which continuation and conjecture is the more plausible?*

Both conjectures predict $\cos \theta < 0$ and therefore $\theta_d > 90°$ (for all d). This shared prediction is encouraging (Problem 6.1); however, being shared means that it does not distinguish between the conjectures.

Does either conjecture match the molecular geometry? An important geometric feature, apart from the bond angle, is the position of the carbon. In one dimension, it lies halfway

H•—1—C—1—•H

between the two hydrogens, so it splits the H–H line segment into two pieces having a 1:1 length ratio.

In two dimensions, the carbon lies on the altitude that connects one hydrogen to the midpoint of the other two hydrogens. The carbon splits the altitude into two pieces having a 1:2 length ratio.

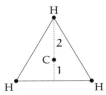

▶ *How does the carbon split the analogous altitude of methane?*

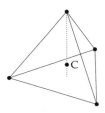

In methane, the analogous altitude runs from the top vertex to the center of the base. The carbon lies at the mean position and therefore at the mean height of the four hydrogens. Because the three base hydrogens have zero height, the mean height of the four hydrogens is $h/4$, where h is the height of the top hydrogen. Thus, in three dimensions, the carbon splits the altitude into two parts having a length ratio of $h/4 : 3h/4$ or $1:3$. In d dimensions, therefore, the carbon probably splits the altitude into two parts having a length ratio of $1:d$ (Problem 6.2).

Because $1:d$ arises naturally in the geometry, $\cos\theta_d$ is more likely to contain $1/d$ rather than $1/2^{d-1}$. Thus, the more likely of the two $\cos\theta_d$ conjectures is that

$$\cos\theta_d = -\frac{1}{d}. \tag{6.1}$$

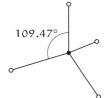

For methane, where $d = 3$, the predicted bond angle is $\arccos(-1/3)$ or approximately $109.47°$. This prediction using reasoning by analogy agrees with experiment and with an honest calculation using analytic geometry (Problem 6.3).

> **Problem 6.2   Carbon's position in higher dimensions**
> Justify conjecture that the carbon splits the altitude into two pieces having a length ratio $1:d$.
>
> **Problem 6.3   Analytic-geometry solution**
> In order to check the solution using analogy, use analytic geometry as follows to find the bond angle. First, assign coordinates (x_n, y_n, z_n) to the n hydrogens, where $n = 1\ldots 4$, and solve for those coordinates. (Use symmetry to make the coordinates as simple as you can.) Then choose two C–H vectors and compute the angle that they subtend.

6.2 Topology: How many regions?

> **Problem 6.4 Extreme case of high dimensionality**
> Draw a picture to explain the small-angle approximation $\arccos x \approx \pi/2 - x$. What is the approximate bond angle in high dimensions (large d)? Can you find an intuitive explanation for the approximate bond angle?

6.2 Topology: How many regions?

The bond angle in methane (Section 6.1) can be calculated directly with analytic geometry (Problem 6.3), so reasoning by analogy does not show its full power. Therefore, try the following problem.

▶ *Into how many regions do five planes divide space?*

This formulation permits degenerate arrangements such as five parallel planes, four planes meeting at a point, or three planes meeting at a line. To eliminate these and other degeneracies, let's place and orient the planes randomly, thereby maximizing the number of regions. The problem is then to find the maximum number of regions produced by five planes.

Five planes are hard to imagine, but the method of easy cases—using fewer planes—might produce a pattern that generalizes to five planes. The easiest case is zero planes: Space remains whole so $R(0) = 1$ (where $R(n)$ denotes the number of regions produced by n planes). The first plane divides space into two halves, giving $R(1) = 2$. To add the second plane, imagine slicing an orange twice to produce four wedges: $R(2) = 4$.

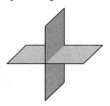

▶ *What pattern(s) appear in the data?*

A reasonable conjecture is that $R(n) = 2^n$. To test it, try the case $n = 3$ by slicing the orange a third time and cutting each of the four pieces into two smaller pieces; thus, $R(3)$ is indeed 8. Perhaps the pattern continues with $R(4) = 16$ and $R(5) = 32$. In the following table for $R(n)$, these two extrapolations are marked in gray to distinguish them from the verified entries.

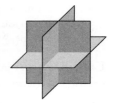

n	0	1	2	3	4	5
R	1	2	4	8	16	32

▶ *How can the $R(n) = 2^n$ conjecture be tested further?*

A direct test by counting regions is difficult because the regions are hard to visualize in three dimensions. An analogous two-dimensional problem would be easier to solve, and its solution may help test the three-dimensional conjecture. A two-dimensional space is partitioned by lines, so the analogous question is the following:

▶ *What is the maximum number of regions into which n lines divide the plane?*

The method of easy cases might suggest a pattern. If the pattern is 2^n, then the $R(n) = 2^n$ conjecture is likely to apply in three dimensions.

▶ *What happens in a few easy cases?*

Zero lines leave the plane whole, giving $R(0) = 1$. The next three cases are as follows (although see Problem 6.5):

R(1)=2   R(2)=4   R(3)=7

> **Problem 6.5 Three lines again**
> The $R(3) = 7$ illustration showed three lines producing seven regions. Here is another example with three lines, also in a random arrangement, but it seems to produce only six regions. Where, if anywhere, is the seventh region? Or is $R(3) = 6$?
>
>
>
> **Problem 6.6 Convexity**
> Must all the regions created by the lines be convex? (A region is convex if and only if a line segment connecting any two points inside the region lies entirely inside the region.) What about the three-dimensional regions created by placing planes in space?

Until $R(3)$ turned out to be 7, the conjecture $R(n) = 2^n$ looked sound. However, before discarding such a simple conjecture, draw a fourth line and carefully count the regions. Four lines make only 11 regions rather than the predicted 16, so the 2^n conjecture is dead.

A new conjecture might arise from seeing the two-dimensional data $R_2(n)$ alongside the three-dimensional data $R_3(n)$.

6.2 Topology: How many regions?

n	0	1	2	3	4
R_2	1	2	4	7	11
R_3	1	2	4	8	

In this table, several entries combine to make nearby entries. For example, $R_2(1)$ and $R_3(1)$—the two entries in the $n = 1$ column—sum to $R_2(2)$ or $R_3(2)$. These two entries in turn sum to the $R_3(3)$ entry. But the table has many small numbers with many ways to combine them; discarding the coincidences requires gathering further data—and the simplest data source is the analogous one-dimensional problem.

▶ *What is the maximum number of segments into which n points divide a line?*

A tempting answer is that n points make n segments. However, an easy case—that one point produces two segments—reduces the temptation. Rather, n points make $n + 1$ segments. That result generates the R_1 row in the following table.

n	0	1	2	3	4	5	n
R_1	1	2	3	4	5	6	$n+1$
R_2	1	2	4	7	11		
R_3	1	2	4	8			

▶ *What patterns are in these data?*

The 2^n conjecture survives partially. In the R_1 row, it fails starting at $n = 2$. In the R_2 row, it fails starting at $n = 3$. Thus in the R_3 row, it probably fails starting at $n = 4$, making the conjectures $R_3(4) = 16$ and $R_3(5) = 32$ improbable. My personal estimate is that, before seeing these failures, the probability of the $R_3(4) = 16$ conjecture was 0.5; but now it falls to at most 0.01. (For more on estimating and updating the probabilities of conjectures, see the important works on plausible reasoning by Corfield [11], Jaynes [21], and Polya [36].)

In better news, the apparent coincidences contain a robust pattern:

n	0	1	2	3	4	5	n
R_1	1	2	3	4	5	6	$n+1$
R_2	1	2	4	7	11		
R_3	1	2	4	8			

▶ *If the pattern continues, into how many regions can five planes divide space?*

According to the pattern,

$$R_3(4) = \underbrace{R_2(3)}_{7} + \underbrace{R_3(3)}_{8} = 15 \tag{6.2}$$

and then

$$R_3(5) = \underbrace{R_2(4)}_{11} + \underbrace{R_3(4)}_{15} = 26. \tag{6.3}$$

Thus, five planes can divide space into a maximum of 26 regions.

This number is hard to deduce by drawing five planes and counting the regions. Furthermore, that brute-force approach would give the value of only $R_3(5)$, whereas easy cases and analogy give a method to compute any entry in the table. They thereby provide enough data to conjecture expressions for $R_2(n)$ (Problem 6.9), for $R_3(n)$ (Problem 6.10), and for the general entry $R_d(n)$ (Problem 6.12).

Problem 6.7 Checking the pattern in two dimensions
The conjectured pattern predicts $R_2(5) = 16$: that five lines can divide the plane into 16 regions. Check the conjecture by drawing five lines and counting the regions.

Problem 6.8 Free data from zero dimensions
Because the one-dimensional problem gave useful data, try the zero-dimensional problem. Extend the pattern for the R_3, R_2, and R_1 rows upward to construct an R_0 row. It gives the number of zero-dimensional regions (points) produced by partitioning a point with n objects (of dimension -1). What is R_0 if the row is to follow the observed pattern? Is that result consistent with the geometric meaning of trying to subdivide a point?

Problem 6.9 General result in two dimensions
The R_0 data fits $R_0(n) = 1$ (Problem 6.8), which is a zeroth-degree polynomial. The R_1 data fits $R_1(n) = n + 1$, which is a first-degree polynomial. Therefore, the R_2 data probably fits a quadratic.

Test this conjecture by fitting the data for $n = 0\ldots 2$ to the general quadratic $An^2 + Bn + C$, repeatedly taking out the big part (Chapter 5) as follows.

a. Guess a reasonable value for the quadratic coefficient A. Then take out (subtract) the big part An^2 and tabulate the leftover, $R_2(n) - An^2$, for $n = 0\ldots 2$.

If the leftover is not linear in n, then a quadratic term remains or too much was removed. In either case, adjust A.

b. Once the quadratic coefficient A is correct, use an analogous procedure to find the linear coefficient B.

c. Similarly solve for the constant coefficient C.

d. Check your quadratic fit against new data ($R_2(n)$ for $n \geqslant 3$).

Problem 6.10   General result in three dimensions
A reasonable conjecture is that the R_3 row matches a cubic (Problem 6.9). Use taking out the big part to fit a cubic to the $n = 0 \ldots 3$ data. Does it produce the conjectured values $R_3(4) = 15$ and $R_3(5) = 26$?

Problem 6.11   Geometric explanation
Find a geometric explanation for the observed pattern. *Hint:* Explain first why the pattern generates the R_2 row from the R_1 row; then generalize the reason to explain the R_3 row.

Problem 6.12   General solution in arbitrary dimension
The pattern connecting neighboring entries of the $R_d(n)$ table is the pattern that generates Pascal's triangle [17]. Because Pascal's triangle produces binomial coefficients, the general expression $R_d(n)$ should contain binomial coefficients.

Therefore, use binomial coefficients to express $R_0(n)$ (Problem 6.8), $R_1(n)$, and $R_2(n)$ (Problem 6.9). Then conjecture a binomial-coefficient form for $R_3(n)$ and $R_d(n)$, checking the result against Problem 6.10.

Problem 6.13   Power-of-2 conjecture
Our first conjecture for the number of regions was $R_d(n) = 2^n$. In three dimensions, it worked until $n = 4$. In d dimensions, show that $R_d(n) = 2^n$ for $n \leqslant d$ (perhaps using the results of Problem 6.12).

6.3 Operators: Euler–MacLaurin summation

The next analogy studies unusual functions. Most functions turn numbers into other numbers, but special kinds of functions—operators—turn functions into other functions. A familiar example is the derivative operator D. It turns the sine function into the cosine function, or the hyperbolic sine function into the hyperbolic cosine function. In operator notation, $D(\sin) = \cos$ and $D(\sinh) = \cosh$; omitting the parentheses gives the less cluttered expression $D \sin = \cos$ and $D \sinh = \cosh$. To understand and learn how to use operators, a fruitful tool is reasoning by analogy: Operators behave much like ordinary functions or even like numbers.

6.3.1 Left shift

Like a number, the derivative operator D can be squared to make D^2 (the second-derivative operator) or to make any integer power of D. Similarly, the derivative operator can be fed to a polynomial. In that usage, an ordinary polynomial such as $P(x) = x^2 + x/10 + 1$ produces the operator polynomial $P(D) = D^2 + D/10 + 1$ (the differential operator for a lightly damped spring–mass system).

How far does the analogy to numbers extend? For example, do $\cosh D$ or $\sin D$ have a meaning? Because these functions can be written using the exponential function, let's investigate the operator exponential e^D.

▶ *What does e^D mean?*

The direct interpretation of e^D is that it turns a function f into e^{Df}.

$$f \longrightarrow \boxed{D} \xrightarrow{Df} \boxed{\exp} \longrightarrow e^{Df}$$

However, this interpretation is needlessly nonlinear. It turns 2f into e^{2Df}, which is the square of e^{Df}, whereas a linear operator that produces e^{Df} from f would produce $2e^{Df}$ from 2f. To get a linear interpretation, use a Taylor series—as if D were a number—to build e^D out of linear operators.

$$e^D = 1 + D + \frac{1}{2}D^2 + \frac{1}{6}D^3 + \cdots . \tag{6.4}$$

▶ *What does this e^D do to simple functions?*

The simplest nonzero function is the constant function $f = 1$. Here is that function being fed to e^D:

$$\underbrace{(1 + D + \cdots)}_{e^D} \underbrace{1}_{f} = 1. \tag{6.5}$$

The next simplest function x turns into $x + 1$.

$$\left(1 + D + \frac{D^2}{2} + \cdots\right) x = x + 1. \tag{6.6}$$

More interestingly, x^2 turns into $(x+1)^2$.

$$\left(1 + D + \frac{D^2}{2} + \frac{D^3}{6} \cdots\right) x^2 = x^2 + 2x + 1 = (x+1)^2. \tag{6.7}$$

6.3 Operators: Euler–MacLaurin summation

> **Problem 6.14 Continue the pattern**
> What is $e^D x^3$ and, in general, $e^D x^n$?

▶ *What does e^D do in general?*

The preceding examples follow the pattern $e^D x^n = (x+1)^n$. Because most functions of x can be expanded in powers of x, and e^D turns each x^n term into $(x+1)^n$, the conclusion is that e^D turns $f(x)$ into $f(x+1)$. Amazingly, e^D is simply L, the left-shift operator.

> **Problem 6.15 Right or left shift**
> Draw a graph to show that $f(x) \to f(x+1)$ is a left rather than a right shift. Apply e^{-D} to a few simple functions to characterize its behavior.
>
> **Problem 6.16 Operating on a harder function**
> Apply the Taylor expansion for e^D to $\sin x$ to show that $e^D \sin x = \sin(x+1)$.
>
> **Problem 6.17 General shift operator**
> If x has dimensions, then the derivative operator $D = d/dx$ is not dimensionless, and e^D is an illegal expression. To make the general expression e^{aD} legal, what must the dimensions of a be? What does e^{aD} do?

6.3.2 Summation

Just as the derivative operator can represent the left-shift operator (as $L = e^D$), the left-shift operator can represent the operation of summation. This operator representation will lead to a powerful method for approximating sums with no closed form.

Summation is analogous to the more familiar operation of integration. Integration occurs in definite and indefinite flavors: Definite integration is equivalent to indefinite integration followed by evaluation at the limits of integration. As an example, here is the definite integration of $f(x) = 2x$.

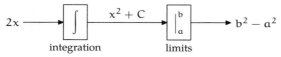

In general, the connection between an input function g and the result of indefinite integration is $DG = g$, where D is the derivative operator and $G = \int g$ is the result of indefinite integration. Thus D and $\int$ are inverses of one another—$D\int = 1$ or $D = 1/\int$—a connection represented by the loop in the diagram. ($\int D \neq 1$ because of a possible integration constant.)

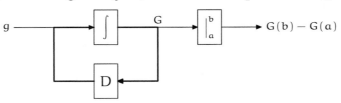

▷ *What is the analogous picture for summation?*

Analogously to integration, define definite summation as indefinite summation and then evaluation at the limits. But apply the analogy with care to avoid an off-by-one or fencepost error (Problem 2.24). The sum $\sum_2^4 f(k)$ includes three rectangles—$f(2)$, $f(3)$, and $f(4)$—whereas the definite integral $\int_2^4 f(k)\,dk$ does not include any of the $f(4)$ rectangle. Rather than rectifying the discrepancy by redefining the familiar operation of integration, interpret indefinite summation to exclude the last rectangle. Then indefinite summation followed by evaluating at the limits a and b produces a sum whose index ranges from a to $b-1$.

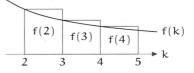

As an example, take $f(k) = k$. Then the indefinite sum $\sum f$ is the function F defined by $F(k) = k(k-1)/2 + C$ (where C is the constant of summation). Evaluating F between 0 and n gives $n(n-1)/2$, which is $\sum_0^{n-1} k$. In the following diagram, these steps are the forward path.

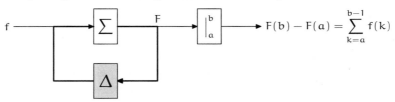

In the reverse path, the new Δ operator inverts Σ just as differentiation inverts integration. Therefore, an operator representation for Δ provides one for Σ. Because Δ and the derivative operator D are analogous, their representations are probably analogous. A derivative is the limit

$$\frac{df}{dx} = \lim_{h \to 0} \frac{f(x+h) - f(x)}{h}. \tag{6.8}$$

6.3 Operators: Euler–MacLaurin summation

The derivative operator D is therefore the operator limit

$$D = \lim_{h \to 0} \frac{L_h - 1}{h}, \tag{6.9}$$

where the L_h operator turns $f(x)$ into $f(x+h)$—that is, L_h left shifts by h.

> **Problem 6.18 Operator limit**
> Explain why $L_h \approx 1 + hD$ for small h. Show therefore that $L = e^D$.

▶ *What is an analogous representation of Δ?*

The operator limit for D uses an infinitesimal left shift; correspondingly, the inverse operation of integration sums rectangles of infinitesimal width. Because summation Σ sums rectangles of unit width, its inverse Δ should use a unit left shift—namely, L_h with $h = 1$. As a reasonable conjecture,

$$\Delta = \lim_{h \to 1} \frac{L_h - 1}{h} = L - 1. \tag{6.10}$$

This Δ—called the finite-difference operator—is constructed to be $1/\Sigma$. If the construction is correct, then $(L-1)\Sigma$ is the identity operator 1. In other words, $(L-1)\Sigma$ should turn functions into themselves.

▶ *How well does this conjecture work in various easy cases?*

To test the conjecture, apply the operator $(L-1)\Sigma$ first to the easy function $g = 1$. Then Σg is a function waiting to be fed an argument, and $(\Sigma g)(k)$ is the result of feeding it k. With that notation, $(\Sigma g)(k) = k + C$. Feeding this function to the $L - 1$ operator reproduces g.

$$[(L-1)\Sigma g](k) = \underbrace{(k+1+C)}_{(L\Sigma g)(k)} - \underbrace{(k+C)}_{(1\Sigma g)(k)} = \underbrace{1}_{g(k)}. \tag{6.11}$$

With the next-easiest function—defined by $g(k) = k$—the indefinite sum $(\Sigma g)(k)$ is $k(k-1)/2 + C$. Passing Σg through $L - 1$ again reproduces g.

$$[(L-1)\Sigma g](k) = \underbrace{\left(\frac{(k+1)k}{2} + C\right)}_{(L\Sigma g)(k)} - \underbrace{\left(\frac{k(k-1)}{2} + C\right)}_{(1\Sigma g)(k)} = \underbrace{k}_{g(k)}. \tag{6.12}$$

In summary, for the test functions $g(k) = 1$ and $g(k) = k$, the operator product $(L-1)\Sigma$ takes g back to itself, so it acts like the identity operator.

This behavior is general—$(L-1)\Sigma 1$ is indeed 1, and $\Sigma = 1/(L-1)$. Because $L = e^D$, we have $\Sigma = 1/(e^D - 1)$. Expanding the right side in a Taylor series gives an amazing representation of the summation operator.

$$\Sigma = \frac{1}{e^D - 1} = \frac{1}{D} - \frac{1}{2} + \frac{D}{12} - \frac{D^3}{720} + \frac{D^5}{30240} - \cdots. \quad (6.13)$$

Because $D\int = 1$, the leading term $1/D$ is integration. Thus, summation is approximately integration—a plausible conclusion indicating that the operator representation is not nonsense.

Applying this operator series to a function f and then evaluating at the limits a and b produces the Euler–MacLaurin summation formula

$$\sum_a^{b-1} f(k) = \int_a^b f(k)\,dk - \frac{f(b) - f(a)}{2} + \frac{f^{(1)}(b) - f^{(1)}(a)}{12} \\ - \frac{f^{(3)}(b) - f^{(3)}(a)}{720} + \frac{f^{(5)}(b) - f^{(5)}(a)}{30240} - \cdots, \quad (6.14)$$

where $f^{(n)}$ indicates the nth derivative of f.

The sum lacks the usual final term $f(b)$. Including this term gives the useful alternative

$$\sum_a^b f(k) = \int_a^b f(k)\,dk + \frac{f(b) + f(a)}{2} + \frac{f^{(1)}(b) - f^{(1)}(a)}{12} \\ - \frac{f^{(3)}(b) - f^{(3)}(a)}{720} + \frac{f^{(5)}(b) - f^{(5)}(a)}{30240} - \cdots. \quad (6.15)$$

As a check, try an easy case: $\sum_0^n k$. Using Euler–MacLaurin summation, $f(k) = k$, $a = 0$, and $b = n$. The integral term then contributes $n^2/2$; the constant term $[f(b) + f(a)]/2$ contributes $n/2$; and later terms vanish. The result is familiar and correct:

$$\sum_0^n k = \frac{n^2}{2} + \frac{n}{2} + 0 = \frac{n(n+1)}{2}. \quad (6.16)$$

A more stringent test of Euler–MacLaurin summation is to approximate $\ln n!$, which is the sum $\sum_1^n \ln k$ (Section 4.5). Therefore, sum $f(k) = \ln k$ between the (inclusive) limits $a = 1$ and $b = n$. The result is

$$\sum_1^n \ln k = \int_1^n \ln k\,dk + \frac{\ln n}{2} + \cdots. \quad (6.17)$$

The integral, from the 1/D operator, contributes the area under the $\ln k$ curve. The correction, from the 1/2 operator, incorporates the triangular protrusions (Problem 6.20). The ellipsis includes the higher-order corrections (Problem 6.21)—hard to evaluate using pictures (Problem 4.32) but simple using Euler–MacLaurin summation (Problem 6.21).

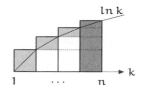

> **Problem 6.19 Integer sums**
> Use Euler–MacLaurin summation to find closed forms for the following sums:
> (a) $\sum_{0}^{n} k^2$ (b) $\sum_{0}^{n} (2k+1)$ (c) $\sum_{0}^{n} k^3$.
>
> **Problem 6.20 Boundary cases**
> In Euler–MacLaurin summation, the constant term is $[f(b) + f(a)]/2$—one-half of the first term plus one-half of the last term. The picture for summing $\ln k$ (Section 4.5) showed that the protrusions are approximately one-half of the last term, namely $\ln n$. What, pictorially, happened to one-half of the first term?
>
> **Problem 6.21 Higher-order terms**
> Approximate $\ln 5!$ using Euler–MacLaurin summation.
>
> **Problem 6.22 Basel sum**
> The Basel sum $\sum_{1}^{\infty} n^{-2}$ may be approximated with pictures (Problem 4.37).
> However, the approximation is too crude to help guess the closed form. As Euler did, use Euler–MacLaurin summation to improve the accuracy until you can confidently guess the closed form. *Hint:* Sum the first few terms explicitly.

6.4 Tangent roots: A daunting transcendental sum

Our farewell example, chosen because its analysis combines diverse street-fighting tools, is a difficult infinite sum.

> Find $S \equiv \sum x_n^{-2}$ where the x_n are the positive solutions of $\tan x = x$.

The solutions to $\tan x = x$ or, equivalently, the roots of $\tan x - x$, are transcendental and have no closed form, yet a closed form is required for almost every summation method. Street-fighting methods will come to our rescue.

6.4.1 Pictures and easy cases

Begin the analysis with a hopefully easy case.

▶ *What is the first root x_1?*

The roots of $\tan x - x$ are given by the intersections of $y = x$ and $y = \tan x$. Surprisingly, no intersection occurs in the branch of $\tan x$ where $0 < x < \pi/2$ (Problem 6.23); the first intersection is just before the asymptote at $x = 3\pi/2$. Thus, $x_1 \approx 3\pi/2$.

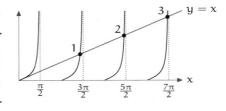

> **Problem 6.23 No intersection with the main branch**
> Show symbolically that $\tan x = x$ has no solution for $0 < x < \pi/2$. (The result looks plausible pictorially but is worth checking in order to draw the picture.)

▶ *Where, approximately, are the subsequent intersections?*

As x grows, the $y = x$ line intersects the $y = \tan x$ graph ever higher and therefore ever closer to the vertical asymptotes. Therefore, make the following asymptote approximation for the big part of x_n:

$$x_n \approx \left(n + \frac{1}{2}\right)\pi. \tag{6.18}$$

6.4.2 Taking out the big part

This approximate, low-entropy expression for x_n gives the big part of S (the zeroth approximation).

$$S \approx \sum \underbrace{\left[\left(n + \frac{1}{2}\right)\pi\right]}_{\approx x_n}^{-2} = \frac{4}{\pi^2} \sum_{1}^{\infty} \frac{1}{(2n+1)^2}. \tag{6.19}$$

The sum $\sum_{1}^{\infty}(2n+1)^{-2}$ is, from a picture (Section 4.5) or from Euler–MacLaurin summation (Section 6.3.2), roughly the following integral.

$$\sum_{1}^{\infty}(2n+1)^{-2} \approx \int_{1}^{\infty}(2n+1)^{-2}\,dn = -\frac{1}{2} \times \frac{1}{2n+1}\bigg|_{1}^{\infty} = \frac{1}{6}. \tag{6.20}$$

6.4 Tangent roots: A daunting transcendental sum

Therefore,

$$S \approx \frac{4}{\pi^2} \times \frac{1}{6} = 0.067547\ldots \qquad (6.21)$$

The shaded protrusions are roughly triangles, and they sum to one-half of the first rectangle. That rectangle has area 1/9, so

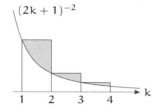

$$\sum_{1}^{\infty} (2n+1)^{-2} \approx \frac{1}{6} + \frac{1}{2} \times \frac{1}{9} = \frac{2}{9}. \qquad (6.22)$$

Therefore, a more accurate estimate of S is

$$S \approx \frac{4}{\pi^2} \times \frac{2}{9} = 0.090063\ldots, \qquad (6.23)$$

which is slightly higher than the first estimate.

▶ *Is the new approximation an overestimate or an underestimate?*

The new approximation is based on two underestimates. First, the asymptote approximation $x_n \approx (n + 0.5)\pi$ overestimates each x_n and therefore underestimates the squared reciprocals in the sum $\sum x_n^{-2}$. Second, after making the asymptote approximation, the pictorial approximation to the sum $\sum_1^\infty (2n+1)^{-2}$ replaces each protrusion with an inscribed triangle and thereby underestimates each protrusion (Problem 6.24).

> **Problem 6.24 Picture for the second underestimate**
> Draw a picture of the underestimate in the pictorial approximation
> $$\sum_{1}^{\infty} \frac{1}{(2n+1)^2} \approx \frac{1}{6} + \frac{1}{2} \times \frac{1}{9}. \qquad (6.24)$$

▶ *How can these two underestimates be remedied?*

The second underestimate (the protrusions) is eliminated by summing $\sum_1^\infty (2n+1)^{-2}$ exactly. The sum is unfamiliar partly because its first term is the fraction 1/9—whose arbitrariness increases the entropy of the sum. Including the $n = 0$ term, which is 1, and the even squared reciprocals $1/(2n)^2$ produces a compact and familiar lower-entropy sum.

$$\sum_{1}^{\infty} \frac{1}{(2n+1)^2} + 1 + \sum_{1}^{\infty} \frac{1}{(2n)^2} = \sum_{1}^{\infty} \frac{1}{n^2}. \qquad (6.25)$$

The final, low-entropy sum is the famous Basel sum (high-entropy results are not often famous). Its value is $B = \pi^2/6$ (Problem 6.22).

▶ *How does knowing $B = \pi^2/6$ help evaluate the original sum $\sum_1^\infty (2n+1)^{-2}$?*

The major modification from the original sum was to include the even squared reciprocals. Their sum is $B/4$.

$$\sum_1^\infty \frac{1}{(2n)^2} = \frac{1}{4} \sum_1^\infty \frac{1}{n^2}. \tag{6.26}$$

The second modification was to include the $n = 0$ term. Thus, to obtain $\sum_1^\infty (2n+1)^{-2}$, adjust the Basel value B by subtracting $B/4$ and then the $n = 0$ term. The result, after substituting $B = \pi^2/6$, is

$$\sum_1^\infty \frac{1}{(2n+1)^2} = B - \frac{1}{4}B - 1 = \frac{\pi^2}{8} - 1. \tag{6.27}$$

This exact sum, based on the asymptote approximation for x_n, produces the following estimate of S.

$$S \approx \frac{4}{\pi^2} \sum_1^\infty \frac{1}{(2n+1)^2} = \frac{4}{\pi^2}\left(\frac{\pi^2}{8} - 1\right). \tag{6.28}$$

Simplifying by expanding the product gives

$$S \approx \frac{1}{2} - \frac{4}{\pi^2} = 0.094715\ldots \tag{6.29}$$

Problem 6.25 Check the earlier reasoning
Check the earlier pictorial reasoning (Problem 6.24) that $1/6 + 1/18 = 2/9$ underestimates $\sum_1^\infty (2n+1)^{-2}$. How accurate was that estimate?

This estimate of S is the third that uses the asymptote approximation $x_n \approx (n + 0.5)\pi$. Assembled together, the estimates are

$$S \approx \begin{cases} 0.067547 & \text{(integral approximation to } \sum_1^\infty (2n+1)^{-2}\text{),} \\ 0.090063 & \text{(integral approximation and triangular overshoots),} \\ 0.094715 & \text{(exact sum of } \sum_1^\infty (2n+1)^{-2}\text{).} \end{cases}$$

Because the third estimate incorporated the exact value of $\sum_1^\infty (2n+1)^{-2}$, any remaining error in the estimate of S must belong to the asymptote approximation itself.

6.4 Tangent roots: A daunting transcendental sum

▶ *For which term of $\sum x_n^{-2}$ is the asymptote approximation most inaccurate?*

As x grows, the graphs of x and $\tan x$ intersect ever closer to the vertical asymptote. Thus, the asymptote approximation makes its largest absolute error when $n = 1$. Because x_1 is the smallest root, the fractional error in x_n is, relative to the absolute error in x_n, even more concentrated at $n = 1$. The fractional error in x_n^{-2}, being -2 times the fractional error in x_n (Section 5.3), is equally concentrated at $n = 1$. Because x_n^{-2} is the largest at $n = 1$, the absolute error in x_n^{-2} (the fractional error times x_n^{-2} itself) is, by far, the largest at $n = 1$.

> **Problem 6.26   Absolute error in the early terms**
> Estimate, as a function of n, the absolute error in x_n^{-2} that is produced by the asymptote approximation.

With the error so concentrated at $n = 1$, the greatest improvement in the estimate of S comes from replacing the approximation $x_1 = (n + 0.5)\pi$ with a more accurate value. A simple numerical approach is successive approximation using the Newton–Raphson method (Problem 4.38). To find a root with this method, make a starting guess x and repeatedly improve it using the replacement

$$x \longrightarrow x - \frac{\tan x - x}{\sec^2 x - 1}. \tag{6.30}$$

When the starting guess for x is slightly below the first asymptote at 1.5π, the procedure rapidly converges to $x_1 = 4.4934\ldots$

Therefore, to improve the estimate $S \approx 0.094715$, which was based on the asymptote approximation, subtract its approximate first term (its big part) and add the corrected first term.

$$S \approx S_{\text{old}} - \frac{1}{(1.5\pi)^2} + \frac{1}{4.4934^2} \approx 0.09921. \tag{6.31}$$

Using the Newton–Raphson method to refine, in addition, the $1/x_2^2$ term gives $S \approx 0.09978$ (Problem 6.27). Therefore, a highly educated guess is

$$S = \frac{1}{10}. \tag{6.32}$$

The infinite sum of unknown transcendental numbers seems to be neither transcendental nor irrational! This simple and surprising rational number deserves a simple explanation.

> **Problem 6.27 Continuing the corrections**
> Choose a small N, say 4. Then use the Newton–Raphson method to compute accurate values of x_n for $n = 1 \ldots N$; and use those values to refine the estimate of S. As you extend the computation to larger values of N, do the refined estimates of S approach our educated guess of $1/10$?

6.4.3 Analogy with polynomials

If only the equation $\tan x - x = 0$ had just a few closed-form solutions! Then the sum S would be easy to compute. That wish is fulfilled by replacing $\tan x - x$ with a polynomial equation with simple roots. The simplest interesting polynomial is the quadratic, so experiment with a simple quadratic—for example, $x^2 - 3x + 2$.

This polynomial has two roots, $x_1 = 1$ and $x_2 = 2$; therefore $\sum x_n^{-2}$, the polynomial-root sum analog of the tangent-root sum, has two terms.

$$\sum x_n^{-2} = \frac{1}{1^2} + \frac{1}{2^2} = \frac{5}{4}. \tag{6.33}$$

This brute-force method for computing the root sum requires a solution to the quadratic equation. However, a method that can transfer to the equation $\tan x - x = 0$, which has no closed-form solution, cannot use the roots themselves. It must use only surface features of the quadratic—namely, its two coefficients 2 and -3. Unfortunately, no plausible method of combining 2 and -3 predicts that $\sum x_n^{-2} = 5/4$.

▶ *Where did the polynomial analogy go wrong?*

The problem is that the quadratic $x^2 - 3x + 2$ is not sufficiently similar to $\tan x - x$. The quadratic has only positive roots; however, $\tan x - x$, an odd function, has symmetric positive and negative roots and has a root at $x = 0$. Indeed, the Taylor series for $\tan x$ is $x + x^3/3 + 2x^5/15 + \cdots$ (Problem 6.28); therefore,

$$\tan x - x = \frac{x^3}{3} + \frac{2x^5}{15} + \cdots. \tag{6.34}$$

The common factor of x^3 means that $\tan x - x$ has a triple root at $x = 0$. An analogous polynomial—here, one with a triple root at $x = 0$, a positive root, and a symmetric negative root—is $(x+2)x^3(x-2)$ or, after expansion, $x^5 - 4x^3$. The sum $\sum x_n^{-2}$ (using the positive root) contains only one term

6.4 Tangent roots: A daunting transcendental sum

and is simply 1/4. This value could plausibly arise as the (negative) ratio of the last two coefficients of the polynomial.

To decide whether that pattern is a coincidence, try a richer polynomial: one with roots at -2, -1, 0 (threefold), 1, and 2. One such polynomial is

$$(x+2)(x+1)x^3(x-1)(x-2) = x^7 - 5x^5 + 4x^3. \tag{6.35}$$

The polynomial-root sum uses only the two positive roots 1 and 2 and is $1/1^2 + 1/2^2$, which is 5/4—the (negative) ratio of the last two coefficients.

As a final test of this pattern, include -3 and 3 among the roots. The resulting polynomial is

$$(x^7 - 5x^5 + 4x^3)(x+3)(x-3) = x^9 - 14x^7 + 49x^5 - 36x^3. \tag{6.36}$$

The polynomial-root sum uses the three positive roots 1, 2, and 3 and is $1/1^2 + 1/2^2 + 1/3^2$, which is 49/36—again the (negative) ratio of the last two coefficients in the expanded polynomial.

▶ *What is the origin of the pattern, and how can it be extended to $\tan x - x$?*

To explain the pattern, tidy the polynomial as follows:

$$x^9 - 14x^7 + 49x^5 - 36x^3 = -36x^3 \left(1 - \frac{49}{36}x^2 + \frac{14}{36}x^4 - \frac{1}{36}x^6\right). \tag{6.37}$$

In this arrangement, the sum 49/36 appears as the negative of the first interesting coefficient. Let's generalize. Placing k roots at $x = 0$ and single roots at $\pm x_1, \pm x_2, \ldots, \pm x_n$ gives the polynomial

$$Ax^k \left(1 - \frac{x^2}{x_1^2}\right)\left(1 - \frac{x^2}{x_2^2}\right)\left(1 - \frac{x^2}{x_3^2}\right)\cdots\left(1 - \frac{x^2}{x_n^2}\right), \tag{6.38}$$

where A is a constant. When expanding the product of the factors in parentheses, the coefficient of the x^2 term in the expansion receives one contribution from each x^2/x_k^2 term in a factor. Thus, the expansion begins

$$Ax^k \left[1 - \left(\frac{1}{x_1^2} + \frac{1}{x_2^2} + \frac{1}{x_3^2} + \cdots + \frac{1}{x_n^2}\right)x^2 + \cdots\right]. \tag{6.39}$$

The coefficient of x^2 in parentheses is $\sum x_n^{-2}$, which is the polynomial analog of the tangent-root sum.

Let's apply this method to $\tan x - x$. Although it is not a polynomial, its Taylor series is like an infinite-degree polynomial. The Taylor series is

$$\frac{x^3}{3} + \frac{2x^5}{15} + \frac{17x^7}{315} + \cdots = \frac{x^3}{3}\left(1 + \frac{2}{5}x^2 + \frac{17}{105}x^4 + \cdots\right). \tag{6.40}$$

The negative of the x^2 coefficient should be $-\sum x_n^{-2}$. For the tangent-sum problem, $\sum x_n^{-2}$ should therefore be $-2/5$. Unfortunately, the sum of positive quantities cannot be negative!

▶ *What went wrong with the analogy?*

One problem is that $\tan x - x$ might have imaginary or complex roots whose squares contribute negative amounts to S. Fortunately, all its roots are real (Problem 6.29). A harder-to-solve problem is that $\tan x - x$ goes to infinity at finite values of x, and does so infinitely often, whereas no polynomial does so even once.

The solution is to construct a function having no infinities but having the same roots as $\tan x - x$. The infinities of $\tan x - x$ occur where $\tan x$ blows up, which is where $\cos x = 0$. To remove the infinities without creating or destroying any roots, multiply $\tan x - x$ by $\cos x$. The polynomial-like function to expand is therefore $\sin x - x \cos x$.

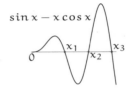

Its Taylor expansion is

$$\underbrace{\left(x - \frac{x^3}{6} + \frac{x^5}{120} - \cdots\right)}_{\sin x} - \underbrace{\left(x - \frac{x^3}{2} + \frac{x^5}{24} - \cdots\right)}_{x \cos x}. \tag{6.41}$$

The difference of the two series is

$$\sin x - x\cos x = \frac{x^3}{3}\left(1 - \frac{1}{10}x^2 + \cdots\right). \tag{6.42}$$

The $x^3/3$ factor indicates the triple root at $x = 0$. And there at last, as the negative of the x^2 coefficient, sits our tangent-root sum $S = 1/10$.

> **Problem 6.28 Taylor series for the tangent**
> Use the Taylor series for $\sin x$ and $\cos x$ to show that
>
> $$\tan x = x + \frac{x^3}{3} + \frac{2x^5}{15} + \cdots. \tag{6.43}$$
>
> *Hint:* Use taking out the big part.

Problem 6.29 Only real roots
Show that all roots of $\tan x - x$ are real.

Problem 6.30 Exact Basel sum
Use the polynomial analogy to evaluate the Basel sum

$$\sum_{1}^{\infty} \frac{1}{n^2}. \tag{6.44}$$

Compare your result with your solution to Problem 6.22.

Problem 6.31 Misleading alternative expansions
Squaring and taking the reciprocal of $\tan x = x$ gives $\cot^2 x = x^{-2}$; equivalently, $\cot^2 x - x^{-2} = 0$. Therefore, if x is a root of $\tan x - x$, it is a root of $\cot^2 x - x^{-2}$. The Taylor expansion of $\cot^2 x - x^{-2}$ is

$$-\frac{2}{3}\left(1 - \frac{1}{10}x^2 - \frac{1}{63}x^4 - \cdots\right). \tag{6.45}$$

Because the coefficient of x^2 is $-1/10$, the tangent-root sum S—for $\cot x = x^{-2}$ and therefore $\tan x = x$—should be $1/10$. As we found experimentally and analytically for $\tan x = x$, the conclusion is correct. However, what is wrong with the reasoning?

Problem 6.32 Fourth powers of the reciprocals
The Taylor series for $\sin x - x \cos x$ continues

$$\frac{x^3}{3}\left(1 - \frac{x^2}{10} + \frac{x^4}{280} - \cdots\right). \tag{6.46}$$

Therefore find $\sum x_n^{-4}$ for the positive roots of $\tan x = x$. Check numerically that your result is plausible.

Problem 6.33 Other source equations for the roots
Find $\sum x_n^{-2}$, where the x_n are the positive roots of $\cos x$.

6.5 Bon voyage

I hope that you have enjoyed incorporating street-fighting methods into your problem-solving toolbox. May you find diverse opportunities to use dimensional analysis, easy cases, lumping, pictorial reasoning, taking out the big part, and analogy. As you apply the tools, you will sharpen them—and even build new tools.

Bibliography

[1] P. Agnoli and G. D'Agostini. Why does the meter beat the second?. *arXiv:physics/0412078v2*, 2005. Accessed 14 September 2009.

[2] John Morgan Allman. *Evolving Brains*. W. H. Freeman, New York, 1999.

[3] Gert Almkvist and Bruce Berndt. Gauss, Landen, Ramanujan, the arithmetic-geometric mean, ellipses, π, and the Ladies Diary. *American Mathematical Monthly*, 95(7):585–608, 1988.

[4] William J. H. Andrewes (Ed.). *The Quest for Longitude: The Proceedings of the Longitude Symposium, Harvard University, Cambridge, Massachusetts, November 4–6, 1993*. Collection of Historical Scientific Instruments, Harvard University, Cambridge, Massachusetts, 1996.

[5] Petr Beckmann. *A History of Pi*. Golem Press, Boulder, Colo., 4th edition, 1977.

[6] Lennart Berggren, Jonathan Borwein and Peter Borwein (Eds.). *Pi, A Source Book*. Springer, New York, 3rd edition, 2004.

[7] John Malcolm Blair. *The Control of Oil*. Pantheon Books, New York, 1976.

[8] Benjamin S. Bloom. The 2 sigma problem: The search for methods of group instruction as effective as one-to-one tutoring. *Educational Researcher*, 13(6):4–16, 1984.

[9] E. Buckingham. On physically similar systems. *Physical Review*, 4(4):345–376, 1914.

[10] Barry Cipra. *Misteaks: And How to Find Them Before the Teacher Does*. AK Peters, Natick, Massachusetts, 3rd edition, 2000.

[11] David Corfield. *Towards a Philosophy of Real Mathematics*. Cambridge University Press, Cambridge, England, 2003.

[12] T. E. Faber. *Fluid Dynamics for Physicists*. Cambridge University Press, Cambridge, England, 1995.

[13] L. P. Fulcher and B. F. Davis. Theoretical and experimental study of the motion of the simple pendulum. *American Journal of Physics*, 44(1):51–55, 1976.

[14] George Gamow. *Thirty Years that Shook Physics: The Story of Quantum Theory*. Dover, New York, 1985.

[15] Simon Gindikin. *Tales of Mathematicians and Physicists*. Springer, New York, 2007.

[16] Fernand Gobet and Herbert A. Simon. The role of recognition processes and look-ahead search in time-constrained expert problem solving: Evidence from grand-master-level chess. *Psychological Science*, 7(1):52-55, 1996.

[17] Ronald L. Graham, Donald E. Knuth and Oren Patashnik. *Concrete Mathematics*. Addison–Wesley, Reading, Massachusetts, 2nd edition, 1994.

[18] Godfrey Harold Hardy, J. E. Littlewood and G. Polya. *Inequalities*. Cambridge University Press, Cambridge, England, 2nd edition, 1988.

[19] William James. *The Principles of Psychology*. Harvard University Press, Cambridge, MA, 1981. Originally published in 1890.

[20] Edwin T. Jaynes. Information theory and statistical mechanics. *Physical Review*, 106(4):620–630, 1957.

[21] Edwin T. Jaynes. *Probability Theory: The Logic of Science*. Cambridge University Press, Cambridge, England, 2003.

[22] A. J. Jerri. The Shannon sampling theorem—Its various extensions and applications: A tutorial review. *Proceedings of the IEEE*, 65(11):1565–1596, 1977.

[23] Louis V. King. On some new formulae for the numerical calculation of the mutual induction of coaxial circles. *Proceedings of the Royal Society of London. Series A, Containing Papers of a Mathematical and Physical Character*, 100(702):60–66, 1921.

[24] Charles Kittel, Walter D. Knight and Malvin A. Ruderman. *Mechanics*, volume 1 of *The Berkeley Physics Course*. McGraw–Hill, New York, 1965.

[25] Anne Marchand. Impunity for multinationals. *ATTAC*, 11 September 2002.

[26] Mars Climate Orbiter Mishap Investigation Board. Phase I report. Technical Report, NASA, 1999.

[27] Michael R. Matthews. *Time for Science Education: How Teaching the History and Philosophy of Pendulum Motion can Contribute to Science Literacy*. Kluwer, New York, 2000.

[28] R.D. Middlebrook. Low-entropy expressions: the key to design-oriented analysis. In *Frontiers in Education Conference, 1991. Twenty-First Annual Conference. 'Engineering Education in a New World Order'. Proceedings*, pages 399–403, Purdue University, West Lafayette, Indiana, September 21–24, 1991.

[29] R. D. Middlebrook. Methods of design-oriented analysis: The quadratic equation revisisted. In *Frontiers in Education, 1992. Proceedings. Twenty-Second Annual Conference*, pages 95–102, Vanderbilt University, November 11–15, 1992.

[30] Paul J. Nahin. *When Least is Best: How Mathematicians Discovered Many Clever Ways to Make Things as Small (or as Large) as Possible*. Princeton University Press, Princeton, New Jersey, 2004.

[31] Roger B. Nelsen. *Proofs without Words: Exercises in Visual Thinking*. Mathematical Association of America, Washington, DC, 1997.

[32] Roger B. Nelsen. *Proofs without Words II: More Exercises in Visual Thinking*. Mathematical Association of America, Washington, DC, 2000.

[33] Robert A. Nelson and M. G. Olsson. The pendulum: Rich physics from a simple system. *American Journal of Physics*, 54(2):112–121, 1986.

[34] R. C. Pankhurst. *Dimensional Analysis and Scale Factors*. Chapman and Hall, London, 1964.

[35] George Polya. *Induction and Analogy in Mathematics*, volume 1 of *Mathematics and Plausible Reasoning*. Princeton University Press, Princeton, New Jersey, 1954.

[36] George Polya. *Patterns of Plausible Inference*, volume 2 of *Mathematics and Plausible Reasoning*. Princeton University Press, Princeton, New Jersey, 1954.

[37] George Polya. *How to Solve It: A New Aspect of the Mathematical Method*. Princeton University Press, Princeton, New Jersey, 1957/2004.

[38] Edward M. Purcell. Life at low Reynolds number. *American Journal of Physics*, 45(1):3–11, 1977.

[39] Gilbert Ryle. *The Concept of Mind*. Hutchinson's University Library, London, 1949.

[40] Carl Sagan. *Contact*. Simon & Schuster, New York, 1985.

[41] E. Salamin. Computation of pi using arithmetic-geometric mean. *Mathematics of Computation*, 30:565–570, 1976.

[42] Dava Sobel. *Longitude: The True Story of a Lone Genius Who Solved the Greatest Scientific Problem of His Time*. Walker and Company, New York, 1995.

[43] Richard M. Stallman and Gerald J. Sussman. Forward reasoning and dependency-directed backtracking in a system for computer-aided circuit analysis. AI Memos 380, MIT, Artificial Intelligence Laboratory, 1976.

[44] Edwin F. Taylor and John Archibald Wheeler. *Spacetime Physics: Introduction to Special Relativity*. W. H. Freeman, New York, 2nd edition, 1992.

[45] Silvanus P. Thompson. *Calculus Made Easy: Being a Very-Simplest Introduction to Those Beautiful Methods of Reasoning Which are Generally Called by the Terrifying Names of the Differential Calculus and the Integral Calculus*. Macmillan, New York, 2nd edition, 1914.

[46] D. J. Tritton. *Physical Fluid Dynamics*. Oxford University Press, New York, 2nd edition, 1988.

[47] US Bureau of the Census. *Statistical Abstracts of the United States: 1992*. Government Printing Office, Washington, DC, 112th edition, 1992.

[48] Max Wertheimer. *Productive Thinking*. Harper, New York, enlarged edition, 1959.

[49] Paul Zeitz. *The Art and Craft of Problem Solving*. Wiley, Hoboken, New Jersey, 2nd edition, 2007.

Index

An italic page number refers to a problem on that page.

ν
  see kinematic viscosity
1 or few
  see few
≈ (approximately equal)   6
π, computing
  arctangent series   64
  Brent–Salamin algorithm   65
∝ (proportional to)   6
∼ (twiddle)   6, 44
ω
  see angular frequency

analogy, reasoning by   99–121
  dividing space with planes   103–107
  generating conjectures
    see conjectures: generating
  operators   107–113
    left shift (L)   108–109
    summation (Σ)   109
  preserving crucial features   100, 118, 120
  pyramid volume   19
  spatial angles   99–103
  tangent-root sum   118–121
  testing conjectures
    see conjectures: testing
  to polynomials   118–121
  transforming dependent variable   101
angles, spatial   99–103
angular frequency   44
Aristotle   xiv
arithmetic–geometric mean   65
arithmetic-mean–geometric-mean inequality   60–66
  applications   63–66
    computing π   64–66
    maxima   63–64
  equality condition   62
  numerical examples   60
  pictorial proof   61–63
  symbolic proof   61
arithmetic mean
  see also geometric mean
  picture for   62
asymptotes of $\tan x$ 114
atmospheric pressure   34

back-of-the-envelope estimates
  correcting   78
  mental multiplication in   77
  minimal accuracy required for   78
  powers of 10 in   78
balancing   *41*
Basel sum ($\sum n^{-2}$)   76, *113*, 116, *121*
beta function   *98*
big part, correcting the
  see also taking out the big part
  additive messier than multiplicative corrections   80
  using multiplicative corrections
    see fractional changes
  using one or few   78
big part, taking out
  see taking out the big part

binomial coefficients   96, *107*
binomial distribution   98
binomial theorem   90, *97*
bisecting a triangle   70–73
bits, CD capacity in   78
blackbody radiation   87
boundary layers   27
brain evolution   57
Buckingham, Edgar   26

calculus, fundamental idea of   31
CD-ROM
   see also CD
   same format as CD   77
CD/CD-ROM, storage capacity   77–79
characteristic magnitudes (typical magnitudes)   44
characteristic times   44
checking units   78
circle
   area from circumference   76
   as polygon with many sides   72
comparisons, nonsense with different dimensions   2
cone free-fall distance   *35*
cone templates   21
conical pendulum   *48*
conjectures
   discarding coincidences   105, 119
   explaining   119
   generating   100, 103, 104, 105
   probabilities of   105
   testing   100, 101, 104, *106*, 111, 119
      getting more data   100, 105, *106*
constants of proportionality
   Stefan–Boltzmann constant   *11*
constraint propagation   5
contradictions   20
convergence, accelerating   65, 68
convexity   *104*
copyright raising book prices   82
Corfield, David   105
cosine
   integral of high power   94–97
   small-angle approximation
      derived   *86*

used   95
cube, bisecting   73

d (differential symbol)   10, 43
degeneracies   103
derivative as a ratio   38
derivatives
   approximating with nonzero Δx 40
   secant approximation   38
      errors in   39
      improved starting point   39
      large error   38
      vertical translation   39
second
   dimensions of   *38*
   secant approximation to   *38*
   significant-change approximation   40–41
      acceleration   43
      Navier–Stokes derivatives   45
      scale and translation invariance   40
   translation invariance   40
desert-island method   32
differential equations
   checking dimensions   42
   linearizing   47, 51–54
   orbital motion   12
   pendulum   46
   simplifying into algebraic equations   43–46
   spring–mass system   42–45
      exact solution   45
      pendulum equation   47
dimensional analysis
   see dimensions, method of; dimensionless groups
dimensionless constants
   Gaussian integral   10
   simple harmonic motion   *48*
   Stefan–Boltzmann law   *11*
dimensionless groups   24
   drag   25
   free-fall speed   24
   pendulum period   48
   spring–mass system   48

dimensionless quantities
  depth of well *94*
  fractional change times exponent *89*
  have lower entropy *94*
  having lower entropy *81*
dimensions
  L for length *5*
  retaining *5*
  T for time *5*
  versus units *2*
dimensions, method of *1–12*
  see also dimensionless groups
  advantages *6*
  checking differential equations *42*
  choosing unspecified dimensions *7, 8–9*
  compared with easy cases *15*
  constraint propagation *5*
  drag *23–26*
  guessing integrals *7–11*
  Kepler's third law *12*
  pendulum *48–49*
  related-rates problems *12*
  robust alternative to solving differential equations *5*
  Stefan–Boltzmann law *11*
dimensions of
  angles *47*
  d (differential) *10*
  dx *10*
  exponents *8*
  integrals *9*
  integration sign $\int$ *9*
  kinematic viscosity ν *22*
  pendulum equation *47*
  second derivative *38*, 43
  spring constant *43*
  summation sign Σ *9*
drag *21–29*
  depth-of-well estimate, effect on *93*
  high Reynolds number *28*
  low Reynolds number *30*
  quantities affecting *23*
drag force
  see drag

e
  in fractional changes *90*
earth
  surface area *79*
  surface temperature *87*
easy cases *13–30*
  adding odd numbers *58*
  beta-function integral *98*
  bisecting a triangle *70*
  bond angles *100*
  checking formulas *13–17*
  compared with dimensions *15*
  ellipse area *16–17*
  ellipse perimeter *65*
  fewer lines *104*
  fewer planes *103*
  guessing integrals *13–16*
  high dimensionality *103*
  high Reynolds number *27*
  large exponents *89*
  low Reynolds number *30*
  of infinite sound speed *92*, *94*
  pendulum
    large amplitude *49–51*
    small amplitude *47–48*
  polynomials *118*
  pyramid volume *19*
  roots of $\tan x = x$ *114*
  simple functions *108*, *112*
  synthesizing formulas *17*
  truncated cone *21*
  truncated pyramid *18–21*
ellipse
  area *17*
  perimeter *65*
elliptical orbit
  eccentricity *87*
  position of sun *87*
energy conservation *50*
energy consumption in driving *82–84*
  effect of longer commuting time *83*
entropy of an expression
  see low-entropy expressions
entropy of mixing *81*
equality, kinds of *6*

estimating derivatives
    see derivatives, secant approximation; derivatives, significant-change approximation
Euler 113
    see also Basel sum
    beta function 98
Euler–MacLaurin summation 112
Evolving Brains 57
exact solution
    invites algebra mistakes 4
examples
    adding odd numbers 58–60
    arithmetic-mean–geometric-mean inequality 60–66
    babies, number of 32–33
    bisecting a triangle 70–73
    bond angle in methane 99–103
    depth of a well 91–94
    derivative of $\cos x$, estimating 40–41
    dividing space with planes 103–107
    drag on falling paper cones 21–29
    ellipse area 16–17
    energy savings from 55 mph speed limit 82–84
    factorial function 36–37
    free fall 3–6
    Gaussian integral using dimensions 7–11
    Gaussian integral using easy cases 13–16
    logarithm series 66–70
    maximizing garden area 63–64
    multiplying 3.15 by 7.21
        using fractional changes 79–80
        using one or few 79
    operators
        left shift (L) 108–109
        summation (Σ) 109–113
    pendulum period 46–54
    power of multinationals 1–3
    rapidly computing 1/13 84–85
    seasonal temperature fluctuations 86–88
    spring–mass differential equation 42–45
    square root of ten 85–86
    storage capacity of a CD-ROM or CD 77–79
    summing $\ln n!$ 73–75
    tangent-root sum 113–121
    trigonometric integral 94–97
    volume of truncated pyramid 17–21
exponential
    decaying, integral of 33
    outruns any polynomial 36
exponents, dimensions of 8
extreme cases
    see easy cases

factorial
    integral representation 36
    Stirling's formula
        Euler–MacLaurin summation 112
        lumping 36–37
        pictures 74
    summation representation 73
    summing logarithm of 73–75
few
    as geometric mean 78
    as invented number 78
    for mental multiplication 78
fractional changes
    cube roots 86
    cubing *83*, *84*
    do not multiply 83
    earth–sun distance 87
    estimating wind power *84*
    exponent of -2 86
    exponent of $1/4$ 87
    general exponents 84–90
    increasing accuracy 85, *86*
    introduced 79–80
    large exponents 89–90, 95
    linear approximation 82
    multiplying 3.15 by 7.21 79
    negative and fractional exponents 86–88
    no plausible alternative to adding 82
    picture 80
    small changes add 82
    square roots 85–86

squaring 82–84
tangent-root sum 117
free fall
   analysis using dimensions 3–6
   depth of well 91–94
   differential equation 4
   impact speed (exact) 4
   with initial velocity 30
fudging 33
fuel efficiency 85

Gaussian integral
   closed form, guessing 14, *16*
   extending limits to ∞ 96
   tail area 55
   trapezoidal approximation 14
   using dimensions 7–11
   using easy cases 13–16
   using lumping 34, 35
GDP, as monetary flow 1
geometric mean
   see also arithmetic mean; arithmetic-mean–geometric-mean theorem
   definition 60
   picture for 61
   three numbers 63
gestalt understanding 59
globalization 1
graphical arguments
   see pictorial proofs

high-entropy expressions
   see also low-entropy expressions
   from quadratic formula 92
How to Solve It xiii
Huygens 48

induction proof 58
information theory 81
integration
   approximating as multiplication
      see lumping
   inverse of differentiation 109
   numerical 14
   operator 109
intensity of solar radiation 86

isoperimetric theorem 73

Jaynes, Edwin Thompson 105
Jeffreys, Harold 26

Kepler's third law *25*
kinematic viscosity (ν) 21, 27

Landau Institute, daunting trigonometric integral from 94
L (dimension of length) 5
Lennard–Jones potential *41*
life expectancy 32
little bit (meaning of d) 10, 43
logarithms
   analyzing fractional changes 90
   integral definition 67
   rational-function approximation 69
low-entropy expressions
   basis of scientific progress 81
   dimensionless quantities are often 81
   fractional changes are often 81
   from successive approximation 93
   high-entropy intermediate steps 81
   introduced 80–82
   reducing mixing entropy 81
   roots of $\tan x = x$ 114
lumping 31–55
   $1/e$ heuristic 34
   atmospheric pressure *34*
   circumscribed rectangle 67
   differential equations 51–54
   estimating derivatives 37–41
   inscribed rectangle 67
   integrals 33–37
   pendulum, moderate amplitudes 51
   population estimates 32–33
   too much 52

Mars Climate Orbiter, crash of 3
Mathematics and Plausible Reasoning xiii
mathematics, power of abstraction 7
maxima and minima *41*, 70
   arithmetic-mean–geometric-mean inequality 63–64

box volume *64*
trigonometry *64*
mental division 33
mental multiplication
   using one or few
      see few
method versus trick 69
mixing entropy 81

Navier–Stokes equations
   difficult to solve 22
   inertial term 45
   statement of 21
   viscous term 46
Newton–Raphson method *76*, 117, *118*
numerical integration 14

odd numbers, sum of 58–60
one or few
   if not accurate enough 79
operators
   derivative (D) 107
      exponential of 108
   finite difference (Δ) 110
   integration 109
   left shift (L) 108–109
   right shift *109*
   summation (Σ) 109–113

parabola, area without calculus *76*
Pascal's triangle *107*
patterns, looking for 90
pendulum
   differential equation 46
   in weaker gravity 52
   period of 46–54
perceptual abilities 58
pictorial proofs 57–76
   adding odd numbers 58–60
   area of circle *76*
   arithmetic-mean–geometric-mean inequality 60–63, *76*
   bisecting a triangle 70–73
   compared to induction proof 58
   dividing space with planes *107*
   factorial 73–75

logarithm series 66–70
Newton–Raphson method *76*
roots of $\tan x = x$ 114
volume of sphere *76*
pictorial reasoning
   depth of well *94*
plausible alternatives
   see low-entropy expressions
Polya, George 105
population, estimating 32
power of multinationals 1–3
powers of ten 78
proportional reasoning 18
pyramid, truncated 17

quadratic formula 91
   high entropy 92
   versus successive approximation 93
quadratic terms
   ignoring 80, 82, 84
   including *85*

range formula *30*
rapid mental division 84–85
rational functions *69*, 101
Re
   see Reynolds number
related-rates problems *12*
rewriting-as-a-ratio trick *68*, *70*, *86*
Reynolds number (Re) 27
   high 27
   low *30*
rigor xiii
rigor mortis xiii
rounding
   to nearest integer 79
   using one or few 78

scale invariance 40
seasonal temperature changes 86–88
seasonal temperature fluctuations
   alternative explanation *88*
secant approximation
   see derivatives, secant approximation
secant line, slope of 38

second derivatives
   see derivatives, second
Shannon–Nyquist sampling theorem
   78
significant-change approximation
   see derivatives, significant-change
   approximation
similar triangles   61, 70
simplifying problems
   see taking out the big part; lumping;
   easy cases; analogy
sine, small-angle approximation
   derived   47
   used   86
small-angle approximation
   cosine   95
   sine   47, 66
solar-radiation intensity   86
space, dividing with planes   103–107
spectroscopy   35
sphere, volume from surface area   76
spring–mass system   42–45
spring constant
   dimensions of   *43*
   Hooke's law, in   42
statistical mechanics   81
Stefan–Boltzmann constant   *11*, 87
Stefan–Boltzmann law
   derivation   *11*
   requires temperature in Kelvin   88
   to compute surface temperature   87
stiffness
   see spring constant
Stirling's formula
   see factorial: Stirling's formula
successive approximation
   see also taking out the big part
   depth of well   92–94
   low-entropy expressions   93
   physical insights   93
   robustness   93
   versus quadratic formula   93
summation
   approximately integration   113, 114
   Euler–MacLaurin   112, *113*
   indefinite   110

integral approximation   74
   operator   109–113
   represented using differentiation   112
   tangent roots   113–121
   triangle correction   74, 113, 115
symbolic reasoning
   brain evolution   57
   seeming like magic   61
symmetry   72

taking out the big part   77–98
   depth of well   92–94
   polynomial extrapolation   *106, 107*
   tangent-root sum   114, 117–118
   trigonometric integral   94–97
Taylor series
   factorial integrand   37
   general   66
   logarithm   66, *69*
      cubic term   *68*
   pendulum period   53
   tangent   118, *120*
L (dimension of length)   5
tetrahedron, regular   99
The Art and Craft of Problem Solving   xiii
thermal expansion   *82*
Thompson, Silvanus   10
thought experiments   18, 50
tools
   see dimensions, method of; easy cases;
   lumping; pictorial proofs; taking out
   the big part; analogy, reasoning by
transformations
   logarithmic   36
   taking cosine   101
trapezoidal approximation   14
tricks
   multiplication by one   85
   rewriting as a ratio   68, *70*, 86
   variable transformation   36, 101
trick versus method   69
tutorial teaching   xiv

under- or overestimate?
   approximating depth of well   92, 93
   computing square roots   86

lumping analysis 54
summation approximation *75*
tangent-root sum 115
using one or few *79*
units
  cancellation 78

Mars Climate Orbiter, crash of 3
  separating from quantities 4
  versus dimensions 2

Wertheimer, Max 59

This book was created entirely with free software and fonts. The text is set in Palatino, designed by Hermann Zapf and available as TeX Gyre Pagella. The mathematics is set in Euler, also designed by Hermann Zapf.

Maxima 5.17.1 and the mpmath Python library aided several calculations.

The source files were created using many versions of GNU Emacs and managed using the Mercurial revision-control system. The figure source files were compiled with MetaPost 1.208 and Asymptote 1.88. The TeX source was compiled to PDF using ConTeXt 2009.10.27 and PDFTeX 1.40.10. The compilations were managed with GNU Make 3.81 and took 10 min on a 2006-vintage laptop. All software was running on Debian GNU/Linux.

I warmly thank the many contributors to the software commons.